FIRE BE█

AND COM█

PROCE█

Raymond Sh█

DELMAR
CENGAGE Learning

Australia • Canada • Mexico • Singapore • Spain • United Kingdom • United States

DELMAR
CENGAGE Learning™

**Fire Behavior and
Combustion Processes**
Raymond Shackelford

Vice President, Career and
Professional Editorial:
Dave Garza

Director of Learning Solutions:
Sandy Clark

Product Development Manager:
Janet Maker

Managing Editor:
Larry Main

Senior Product Manager:
Jennifer Starr

Editorial Assistant:
Maria Conto

Vice President, Career
and Professional Marketing:
Jennifer McAvey

Marketing Director:
Deborah S. Yarnell

Marketing Manager:
Erin Coffin

Marketing Coordinator:
Shanna Gibbs

Production Director:
Wendy Troeger

Production Manager:
Mark Bernard

Senior Art Director:
Bethany Casey

Technology Project Manager:
Christopher Catalina

Production Technology Analyst:
Thomas Stover

Library of Congress Control Number: 2008936421
ISBN-13: 978-1-4018-8016-3
ISBN-10: 1-4018-8016-9

Delmar
5 Maxwell Drive
Clifton Park, NY 12065-2919
USA

Cengage Learning products are represented in Canada by Nelson Education,
Ltd.

For your lifelong learning solutions, visit **delmar.cengage.com**

Visit our corporate website at **cengage.com.**

Notice to the Reader

Publisher does not warrant or guarantee any of the products described
herein or perform any independent analysis in connection with any of the
product information contained herein. Publisher does not assume, and expressly disclaims, any obligation to obtain and include information other
than that provided to it by the manufacturer. The reader is expressly warned
to consider and adopt all safety precautions that might be indicated by the
activities described herein and to avoid all potential hazards. By following
the instructions contained herein, the reader willingly assumes all risks in
connection with such instructions. The publisher makes no representations
or warranties of any kind, including but not limited to, the warranties of fit-
ness for particular purpose or merchantability, nor are any such representations implied with respect to the material set forth herein, and the publisher
takes no responsibility with respect to such material. The publisher shall not
be liable for any special, consequential, or exemplary damages resulting, in
whole or part, from the readers' use of, or reliance upon, this material.

Printed in Canada
1 2 3 4 5 XX 10 09 08

CONTENTS

PREFACE

ABOUT THIS BOOK

Fire Behavior and Combustion Processes was designed to provide a straight-forward yet comprehensive resource for students enrolled in fire science degree programs, or as a refresher for active firefighters. It provides an understanding of the basic principles of fire chemistry, the processes of fire combustion, and fire behavior. The subject of fire behavior is often a complex one, and this book seeks to clarify theoretical concepts, explain their importance, and illustrate how they can be applied in a practical way when responding to emergency situations. There is also a special emphasis on safety, with each explanation drawing a connection between how a fire behaves and how it affects the safety of individual firefighters and their team.

In addition, this textbook covers all the objectives and content for the Fire and Emergency Services Higher Education (FESHE) course, *Fire Behavior and Combustion*, instituted by the National Fire Academy. FESHE is an organization of the secondary educational organizations that meet at the National Fire Academy annually to develop and approve a national curriculum that is standardized in scope, content, and outcomes. A correlation guide, which cross references the FESHE course objectives with the content in this book, can be found on page xi.

HOW TO USE THIS BOOK

This book takes the complex layers of fire behavior and separates them out to help facilitate learning. Chapters 1 through 4 provide the background and theory involved in fire behavior. They introduce the student to the history and foundations of fire protection, fire chemistry, fire combustion processes, and fire extinguishment. Chapters 5 through 10 proceed to apply the theory to various incidents likely to be encountered by firefighters, and how firefighters can react safely for a successful outcome.

Chapter 1 sets up context for all the chapters that follow by providing a look back at pivotal fires in history and important lessons learned. This chapter also includes a section on firefighter injury and fatalities in the United States, stressing the importance of safe practices in the fire service.

Chapter 2 highlights the basics of fire chemistry, including an explanation of basic physical and chemical properties of fire and how it relates to firefighter response.

Chapter 3 explains the combustion process, covering fire classifications, underlying causes, the various stages, and types of fire events.

Chapter 4 explores extinguishing agents that are available to firefighters and their various applications. A section that identifies the latest technological advances in fire extinguishment, such as compressed air foam and ultrafine water mist, keeps firefighters knowledgeable of systems they may encounter on the job.

Chapter 5 provides the foundation for the chapters that follow by introducing the concepts of fire fighting strategies and tactics. In this chapter, the book begins to explore the various types of attack modes based on different types of fire scenarios.

Chapter 6 moves deeper into the discussion of fire fighting strategies and tactics by presenting special concerns in fire fighting activities. It includes a discussion of specific actions: prefire and postfire event planning, addressing utility concerns, working in confined enclosures, and salvage and overhaul considerations.

Chapter 7 focuses specifically on high-rise building fires and the special actions and considerations involved in response to this fire scenario. This chapter clearly distinguishes high-rise fires from others and explains the unique challenges associated with fighting these types of fires.

Chapter 8 covers wildland fire fighting for those areas of the country where wildland fires are a concern. Included is a section on special techniques for wildland fire response, as well as the various tools and resources required for fighting these fires. This offers students an opportunity to expand their knowledge of fire fighting in different types of locales.

Chapter 9 proceeds to cover transportation fires and all related fire fighting activities that are a consideration for response to these types of fires. It includes a look at fires in passenger vehicles, motor homes, buses, recreational vehicles, trucks, railcars, aircraft, and boats. The chapter also considers the latest technologies, such as hybrid and hydrogen-powered vehicles, to ensure that firefighters have an understanding of the unique challenges associated with vehicle fire response today.

Chapter 10 concludes the book with a concentrated look at hazardous materials incidents and terrorist events that are a consideration for firefighters of today. A discussion of the National Incident Management System (NIMS) is also included to ensure that firefighters have an understanding of the logistics of this emergency response plan.

A recommended approach to using the text would be to carefully read Chapters 1 through 4 and answer all of the review questions at the end of each chapter before tackling the remaining chapters of the book. Once completed, the student should apply the basic chemistry and combustion principles found in the earlier chapters. For example, Chapter 5 applies the basic chemistry and combustion as well as the extinguishing principles to various occupancies that fire responders will encounter.

FEATURES OF THIS BOOK

This book features many valuable tools to assist in learning and applying the information presented in the chapters.

- A **FESHE Correlation Grid** at the front of the book correlates the content in the book to the National Fire Academy FESHE course outcomes to ensure standardized training.
- *Safety* is emphasized throughout the book, including important *Safety* notes to highlight recommended safe practices to avoid dangerous situations and ensure the safety of the individual firefighter and the team.
- *Notes* are integrated in the chapters, and reinforce important concepts and applications to ensure a complete understanding of fire behavior and provide a helpful resource for review.
- *Key Terms* highlighted in the chapters are also listed at the end of the chapters with complete definitions to provide another helpful resource for review.
- *Technical References* are included in Appendices A and B to provide a handy tool for review. Included among the references are conversion tables, formulas, and standard fire orders.

SUPPLEMENT TO THIS BOOK

Delmar Cengage Learning is proud to offer the following supplemental materials to assist students and instructors in learning and delivering the content presented in this book.

The **Instructor's Guide on CD-ROM** offers the following tools for instructors.

- **Lesson Plans** with correlating *PowerPoint* presentations outline the important information in each chapter to provide a guide for classroom instruction.
- *PowerPoint* presentations follow the outline of the *Lesson Plans* and combine bulleted information with photos and graphics to reinforce important points.
- *Computerized Testbanks* in ExamView 6.0 offer instructors an option for evaluating student knowledge of the concepts presented in each chapter. Instructors may use the exams as they are written, or edit the questions and create new exams based on the needs of the class.
- A *Workbook* is available to offer students additional questions to practice their knowledge of theory and applications presented in each chapter.
- A *FESHE Correlation Grid* correlates chapter content with the outcomes outlined in the National Fire Academy FESHE course, *Fire Behavior and Combustions,* to ensure that the instructor complies with standardized training.

Please visit us at www.delmarfire.cengage.com to discover other fire and FESHE-related titles available.

ABOUT THE AUTHOR

Dr. Shackelford has been a member of the fire service for over forty years and an educator for a quarter of a century. He began his public service career as a firefighter with the Sierra Madre Volunteer Fire Department in 1960 and served with the San Marino Fire Department prior to joining the Los Angeles County Fire Department in July 1964, where he served for twenty-nine years, promoting to the rank of Deputy Fire Chief. He served as inaugural Fire Chief of the newly formed Chino Valley Independent Fire District from 1990 through 1995.

A visionary, Dr. Shackelford has been instrumental in the shape of the fire service, as a senior fire official, author, lecturer, and professor and director of the Fire Protection Administration Program at California State University, Los Angeles since 1995.

In addition to teaching and mentoring thousands of firefighters and fire administrators, Dr. Shackelford has led and participated in numerous committees and projects that are seen in current academic thought, and in fire department emergency and administrative management practices, including the use of strategic planning, modern budgeting and personnel management practices, numerous public education and fire prevention programs, and most notably, in the development of FIRESCOPE and the Incident Command System.

Dr. Shackelford has served in numerous other positions, a sample of which include Board Member for the Los Angeles County Firefighter's IAFF Local #1014, President of the Los Angeles County and the San Bernardino County Fire Chiefs Associations, and served on the California Fire Chiefs Association Board of Directors.

Dr. Shackelford has earned an Associate of Science Degree in Electronics, an Associate of Science Degree in Fire Technology, a Bachelor of Science Degree in Public Administration, and a Master of Science Degree in Public Administration. He earned a Doctorate Degree in Public Administration in 2002 at the University of La Verne. His dissertation compared and contrasted fifty different fire departments across the United States, analyzing, comparing, and contrasting the impact of the fire accreditation process.

His background and knowledge put him in great demand by governmental agencies and nonprofit organizations throughout the United States and as a consultant to governments around the world. He has visited Japan, China, and Taiwan a number of times and has provided valuable assistance in emergency preparedness and fire service operations.

Technical Edit: Todd Haines

Todd Haines' interest in fire safety began 18 yrs ago when he joined his first volunteer fire dept in Wethersfield, CT. Since then, Todd has earned both his BS in Fire Protection & Safety Engineering and MS in Fire and Emergency Management from Oklahoma State University and maintained volunteer fire service in Oklahoma and Texas. His career highlights include: Fire Protection Engineer at Dallas/Fort Worth International Airport; Hazardous Materials Specialist for the Federal Bureau of Investigation, Hazardous Materials Response Unit; Engineering Associate with the Austin Fire Department—Special Operations and Hazardous Materials Engineering Divisions and Safety Engineering in the Aerospace and Petrochemical Industry.

ACKNOWLEDGMENTS

The list of persons who provided assistance and feedback for this book is numerous. First, to the editors at Delmar Cengage Learning for all of their support and assistance, I wish to extend my gratitude. They have extended tremendous patience in helping and encouraging the work needed to put this text together.

Also, special thanks to all of my colleagues attending and participating in the annual FESHE conference who I consulted for ideas and thoughts on what should be included in this text. Everyone has been helpful in providing the needed ideas and encouragement to finish this text.

And to the following reviewers who participated in the development of the book and provided helpful comments as we finalized the manuscript, I wish to offer my appreciation.

Gail Hughes
Fire Science Coordinator
University of Alaska
Anchorage, Alaska

Albert Iannone
Director of Fire Technology
American River College
Sacramento, California

Steve Malley
Director, Weatherford College
Regional
Fire Academy
Weatherford, Texas

M.B. Oliver
Director of Fire Science
Midland College
Midland, Texas

Ken Staelgraeve
Director of Fire Science
Macomb Community College
Warren, Michigan

FEEDBACK

The author requests your suggestions on how the text can be improved. Questions and comments can be directed to the author via e-mail at rshacke@calstatela.edu.

FIRE AND EMERGENCY SERVICE HIGHER EDUCATION (FESHE)

In June 2001, the U.S. Fire Administration hosted the third annual Fire and Emergency Services Higher Education Conference at the National Fire Academy campus in Emmitsburg, Maryland. Attendees from state and local fire service training agencies, as well as colleges and universities with fire-related degree programs, attended the conference and participated in work groups. Among the significant outcomes of the working groups was the development of standard titles, outcomes, and descriptions for six core associate-level courses for the model fire science curriculum that had been developed by the group the previous year. The six core courses are *Fundamentals of Fire Protection, Fire Protection Systems, Fire Behavior and Combustion, Fire Protection Hydraulics and Water Supply, Building Construction for Fire Protection,* and *Fire Prevention.*

FIRE BEHAVIOR AND COMBUSTION COURSE CONTENT

The National Fire Science Curriculum Advisory Committee identified ten desired outcomes, involving eleven content areas for the course. This text addresses each desired outcome within the eleven content areas.

Fire Behavior and Combustion Core Course—Desired Outcomes

1. Identify physical properties of the three states of matter.
2. Categorize the components of fire.
3. Recall the physical and chemical properties of fire.
4. Describe and apply the process of burning.
5. Define and use basic terms and concepts associated with the chemistry and dynamics of fire.
6. Describe the dynamics of fire.
7. Discuss various materials and their relationship to fires as fuel.
8. Demonstrate knowledge of the characteristics of water as a fire suppression agent.
9. Articulate other suppression agents and strategies.
10. Compare other methods and techniques of fire extinguishments.

FESHE Content Area Comparison

The following table provides a comparison of the eleven FESHE content areas in this text.

FIRE AND EMERGENCY SERVICES HIGHER EDUCATION (FESHE) COURSE CORRELATION GRID

Name:	Fire Behavior and Combustion	*Fire Behavior and Combustion* Chapter Reference
Course Description:	This course explores the theories and fundamentals of how and why fires start, spread, and how they are controlled.	
Prerequisite:	*None.*	
Course Outline:	I. Introduction	
	A. Matter and Energy	1,2
	B. The Atom and its Parts	2
	C. Chemical Symbols	2
	D. Molecules	2
	E. Energy and Work	1
	F. Forms of Energy	3
	G. Transformation of Energy	3
	H. Laws of Energy	3
	II. Units of Measurements	
	A. International (SI) Systems of Measurement	1, Appendix A
	B. English Units of Measurement	1, Appendix A
	III. Chemical Reactions	
	A. Physical States of Matter	2
	B. Compounds and Mixtures	2
	C. Solutions and Solvents	2
	D. Process of Reactions	2
	IV. Fire and the Physical World	
	A. Characteristics of Fire	3
	B. Characteristics of Solids	2
	C. Characteristics of Liquids	2
	D. Characteristics of Gases	2
	V. Heat and its Effects	
	A. Production and Measurement of Heat	1,2,4
	B. Different Kinds of Heat	3

Course Outline:		
	VI. Properties of Solids Materials	
	A. Common Combustible Solids	3
	B. Plastic and Polymers	2
	C. Combustible Metals	3,4
	D. Combustible Dust	2
	VII. Common Flammable Liquids and Gases	
	A. General Properties of Gases	2
	B. The Gas Laws	2
	C. Classification of Gases	2
	D. Compressed Gases	2,4
	VIII. Fire Behavior	
	A. Stages of Fire	3
	B. Fire Phenomena	3
	1. Flashover	3
	2. Backdraft	3
	3. Rollover	3
	4. Flameover	3
	C. Fire Plumes	3
	IX. Fire Extinguishment	
	A. The Combustion Process	3
	B. The Character of Flame	3
	C. Fire Extinguishment	3,4
	X. Extinguishing Agents	
	A. Water	4
	B. Foams and Wetting Agents	4
	C. Inert Gas Extinguishing Agents	4
	D. Halogenated Extinguishing Agents	1
	E. Dry Chemical Extinguishing Agents	4
	F. Dry Powder Extinguishing Agents	4
	XI. Hazards by Classification Types	
	A. Hazards of Explosives	10
	B. Hazards of Compressed and Liquefied Gases	4
	C. Hazards of Flammable and Combustible Liquids	3,4
	D. Hazards of Flammable Solids	3,4
	E. Hazards of Oxidizing Agents	4
	F. Hazards of Poisons	10
	G. Hazards of Radioactive Substances	10
	H. Hazards of Corrosives	10

DEATH AND INJURY RATES: THE AMERICAN EXPERIENCE

For a number of years, Americans have experienced a high annual civilian fire death and injury rate in comparison to other industrialized countries. Researchers have identified a number of reasons for this high rate and have recommended an increase in emphasis on fire prevention and public education activities (Schaenman, 1982; Schaenman & Seits, 1985; Geneva Association newsletter). Unfortunately, these researchers found other serious concerns as well.

In addition to the high civilian fire death rate, the United States also has a high firefighter death rate. According to the 2002 FEMA "Firefighter Fatality Retrospective Study," one hundred firefighters die annually and many more are injured. This average can be traced back to 1977, and is fairly consistent, although in the past few years, the annual death rate trend has slowly increased. This trend has prompted the U.S. Fire Administration (USFA) to commit to reducing fire service deaths by 25% over the next five to ten years. The following table shows the latest available averages for fire deaths, injuries, and total injuries.

Firefighter fatalities, fire ground injuries, and total injuries (1995–2004)

Year	Deaths[1]	Fire Ground Injuries[2]	Total Injuries
1998	93	43,080	87,500
1999	113	45,550	88,500
2000	105	43,065	84,550
2001	105[3]	41,395	82,250
2002	100	37,860	80,800
2003	113	38,045	78,750
2004	119[4]	36,880	75,840
2005	115	41,950	80,100
2006	106	44,210	83,400
2007	118	N/A	N/A

[1] This figure reflects the number of deaths as published in USFA's annual report on firefighter fatalities.
[2] This figure reflects the number of injuries as published in NFPA's annual report on firefighter injuries.
[3] In 2001, an additional 341 FDNY firefighters, three fire safety directors, and two FDNY paramedics died in the line of duty at the World Trade Center on 9/11.
[4] The Hometown Heroes Survivors Benefit Act of 2003 has resulted in an approximate 10% increase to the total number of firefighter fatalities counted for the annual USFA report on firefighter fatalities in the United States beginning with CY 2004.

Source: USFA, available at http://www.usfa.dhs.gov/fireservice/fatalities/statistics/casualties.shtm.

The report identified the four leading causes of firefighter deaths. The leading cause of death (44%) for on-duty firefighters is heart attack, followed by trauma such as a fatal head or internal injury (27%). The third highest cause of death is asphyxiation and burns (20%). The fourth area of firefighter deaths is categorized as miscellaneous. This category includes deaths and injuries that range from responding from and returning to quarters, training activities, station maintenance work, and other activities. This area is responsible for 9% of the deaths.

ADDRESSING THE ISSUE OF FIREFIGHTER FATALITIES

The 2002 FEMA retrospective firefighter fatality report concluded that the high percentage rate of firefighter deaths can be reduced by implementing the following:

1. preemployment medical examinations for existing heart problems;
2. regular physical exercise;
3. dietary management; and
4. required annual physical check-ups.

It is estimated that if these four recommendations were implemented nationwide, 25% to 50% of firefighter heart attacks could be prevented.

The trauma injuries including internal and head injuries (27%) is the second highest category of deaths. Firefighters included in this category are generally less than thirty-five years old. The studies noted that although it may not be possible to reduce all deaths and injuries, all results indicated the total number could be reduced by researching past fires where deaths have occurred, then identifying the specific cause of death and implementing the changes needed to prevent a reoccurrence.

The third category includes asphyxiation and burns (20%). Many of these deaths are attributed to the need for additional training, better equipment, and more widespread sharing of information from postfire conferences.

In the miscellaneous category, 9% of total deaths occurred. Many of the deaths and injuries in this category resulted from responding to and returning from quarters or during training activities and station maintenance work. Most of these accidents can be prevented using driver training courses, training drills and safety reminders, and a strongly enforced safety policy.

THE IMPORTANCE OF PLANNING

Firefighters and fire officers know that the ability to be successful in regard to fire attack comes from long hours of study, careful fire preplanning, and many years of firefighting experience. These are all combined with a carefully planned postfire review where the positive and the negative procedures are examined in an effort to benefit from the lessons learned. These preplanning and postincident analysis conferences should be planned to keep a continual learning and improvement cycle in place within the fire department (see Chapter 6).

The identification and isolation of fire fighting problems is of little use if firefighters do not follow up on information gained from the postfire planning session. This session should be designed by the officer to allow positive and corrective comments to be fully discussed so lessons can be learned to improve actions on future incidents.

REFERENCE

FEMA, USFA. (2002). *Firefighter Fatality Retrospective Study*, http://www.usfa.dhs.gov/downloads/pdf/publications/fa-220.pdf

USFA. (2007). *Firefighter Death and Injury Report*, http://www.usfa.dhs.gov/fireservice/fatalities/statistics/casualties.shtm.

Chapter 1

AMERICAN FIRE SERVICE: THE PAST, PRESENT, AND FUTURE

Learning Objectives

Upon completion of this chapter, you should be able to:

- Examine how the history of our society has shaped the American attitude toward fire prevention and fire control efforts.
- Describe and explain how other countries approach the control of fires and compare and contrast their approach with U.S. efforts.
- Describe new technologies and systems the fire service has implemented in recent years.
- Describe the fire service of today, its successes, its problems, and its efforts toward improvement.
- Examine and envision the challenges and opportunities open to the fire service in the twenty-first century.

INTRODUCTION

The culture in the United States and its rich and complex history have shaped the American fire service into what it is today. By reviewing the past and comparing it to the present-day fire service, we can conclude that although much has been done to reduce fire deaths and losses, improvement is still needed. Further reducing these casualties and losses remains the biggest challenge facing the fire service as society grows and changes.

To fully understand the fire service of today and envision its future, it is important to reflect upon its history to see how it has evolved. Figure 1-1 illustrates the American firefighters of the past. The objectives of this chapter include a brief synopsis of the U.S. fire service's complicated history, an examination of what the fire service is currently doing, and a look at its future challenges.

FIRE SERVICE OF THE PAST

Conflagration
A fire with major building-to-building flame spread over a great distance.

Since the early days of colonial settlements in New America, fire has been a big part of history. These early settlements experienced several major **conflagrations** and a number of large fires. In some cases the settlements were totally consumed by fire. This pattern of occasional conflagrations and large fires has continued through history to present day.

Figure 1-1 *The fire service of the past, present, and future has a tradition of many service challenges.*

The Industrial Revolution

The last quarter of the nineteenth century found the United States undergoing an industrial revolution. The cities were expanding and becoming crowded as more people found employment in factories and took up residence in the cities rather than working and living on farms. This increased growth in building construction led to poorly constructed and unplanned developments in the cities. Many structures were sizeable in comparison to the small structures of earlier settlements. While minimum building and fire codes were enacted, they were not strictly enforced and in many cases fire resistance and fire safety issues were ignored.

Documenting this period of the American fire service, the editor of *Fire Engineering* magazine comments, "Little control was exerted over construction methods and materials, fire loading and occupancy loads . . . and while there were many chances for the fire service to step up to the challenge there is little evidence that fire departments were quick to seize the opportunity." This occurred because at the time, building laws were under the jurisdiction of the building departments. Building inspectors were employees of either the state or city building departments. Those agencies were often corrupt or simply failed to see the importance of safe buildings. Buildings that could be quickly and cheaply constructed were of greatest importance.

Corruption in building departments resulted in poor construction practices and little or no fire code enforcement. This eventually resulted in large fires that destroyed entire cities. Some fires resulted in the loss of civilians as well as firefighters. The Industrial Revolution proved to be very challenging to the fire service because growth and expansion took precedence over fire and life safety concerns.

Two Great Fires On October 9, 1871, two major fires raged out of control on the same day in the United States. One was a forest fire in northeastern Wisconsin, which resulted in the death of 1,152 persons and burned 2,400 square miles of forested land. The second fire, known as the Great Chicago Fire, was a much more widely publicized fire. It attracted media attention and overshadowed the Wisconsin forest fire even though there were fewer lives lost.

Because of the media attention, the Chicago fire was widely celebrated in many songs, fables, and tall tales in U.S. history. The fire was rumored to have been started by a cow kicking over a lantern, but there is now evidence that leads investigators to believe that strong winds coupled with an earlier lumberyard fire resulted in a flying ember that started a fire in the straw in the famous Mrs. O'Leary's cow barn. (Learn more about the fire at http://egov.cityofchicago.org/webportal/COCWebPortal/COC_EDITORIAL/HistoryOfTheChicagoFireDepartment_1.pdf.)

It is reported that the Chicago fire resulted in the death of 300 persons and destroyed buildings in a 2.5 square mile area. It continued to burn for two days before the firefighters were able to stop its advance. Quick, poor-quality construction using little or no fire-resistant materials was a large factor in the widespread devastation.

After the Great Chicago Fire, the city council approved an ordinance that required new buildings to be "built of stone or of masonry." Even though it appeared that the city wanted to fix the problems that caused the fire, within a very short time, this new ordinance was forgotten. As a result, the fire insurance companies became concerned of the possibility of another conflagration.

Insurance Companies Respond In 1874, the fire insurance companies in the United States were being represented by a newly formed entity. The National Board of Fire Underwriters was organized in an effort to bring the importance of enforcement of the ordinances created in response to the catastrophe to the attention of the Chicago City Council. However, the city council did not respond to the repeated requests and the insurance companies retaliated by closing all insurance offices. The city council then agreed to comply and the insurance companies responded by reopening the offices and increasing the fire insurance rates by 20%.

Insurance Services Office (ISO)

An agency funded by the insurance companies to independently apply the grading schedule to cities (fire departments) and set the rate for fire insurance premiums.

The actions resulting from the Great Chicago Fire laid the groundwork for today's **Insurance Services Office (ISO)** rating schedule for fire protection services. A consistent and uniform method was needed to assess the city's and the fire department's ability to prevent a conflagration. The control is higher fire insurance premiums for a poor fire rating. Many components within the rating schedule have changed over the years, but the ISO still measures the fire department's overall ability to control large fires and sets the rate for fire insurance premiums for individual structures.

The Decade of Conflagrations

The years 1900 to 1909 have been called "the decade of conflagrations." Five of the most significant fires in the United States occurred during this period. These fires brought the public's attention to areas that had been contributing to the devastation and loss of lives and property for a long time. Areas of particular concern were:

- the need for fire-resistive construction materials in buildings;
- a need for a dependable water supply;
- the regular inspection of fire and life-saving equipment; and
- the safe storage of combustible and flammable materials.

A fire and earthquake brought the need for fire-resistive construction to the forefront of concern in 1906. The San Francisco, California, earthquake and fire was described as a terrible "double scourge," meaning the earthquake broke gas lines and water mains which caused fires. As a result, an estimated 674 people died and 3,500 others were injured. The earthquake and fire destroyed 28,000 buildings occupying 514 city blocks. As a result, the city of San Francisco installed a special fire water system so that the domestic water system is backed up in the event of an earthquake. This "decade of conflagrations" made it clear that serious fire problems existed.

The California earthquake and fire also triggered the formation of the National Fire Protection Association (NFPA) and, as mentioned, the ISO grading schedule. The NFPA began by developing recommendations for fire and safety standards, and the ISO started grading and developing a rating system for fire services.

Despite the growing awareness of the fire problems, fire service personnel have made only limited progress in the lines of safety. It is incumbent upon those new to the fire service to learn from the past and become proactive to prevent another "decade of conflagrations" from occurring.

This cyclical occurrence of conflagrations and large fires and then rebuilding has continued to exist since the early 1900s. It is like a tradition in the United States. At present, we are still plagued with conflagrations that devastate our towns and cities. For today's fire students, a better understanding of fire combustion processes, the use of improved fire-resistant building materials and methods, and the use of improvements in fire technologies can stop this cyclical reoccurrence of conflagrations.

FIRE SERVICE OF TODAY

Discovering the U.S. Fire Problem

During the early 1960s, the International Association of City/County Managers joined with the International Association of Fire Chiefs to conduct and publish a number of special studies to determine and communicate the extent of the fire problem in the United States. These studies all pointed to the need to examine the U.S. fire problem in depth. They found that despite U.S. spending of billions of dollars for fire protection, communities continued to experience a large loss in life and property. As these reports surfaced, the U.S. fire service began to recognize the need to understand what was causing the loss and how to change the trend.

As a result of these reports, the U.S. fire service began to recognize the need to include public support in finding a solution to combating and controlling the nation's fire problems. In 1968, concerned fire service leaders called for a conference to bring together the interested stakeholders and identify the fire problem. At the conference, held in Racine, Wisconsin, at the Wingspread Conference Center, representatives of organizations concerned with the nationwide fire problem assembled to find resolutions to the nation's fire problems. Their report encouraged the need for a comprehensive examination of the fire problem

America Burning
The 1973 report from a committee appointed by the president to investigate and report back its findings regarding U.S. fire service.

in the United States. The result was the 1973 presidential committee report, ***America Burning.***

The *America Burning* report identified a number of areas for improvement in the fire service. Recommendations included improved fire code enforcement, data collection and processing, as well as the need for improving the quality of training, equipment, technology, and education for the fire service. Over the next few years, some of these recommendations were addressed, but the progress did not meet expectations.

However, two of the recommendations that were addressed resulted in the establishment of the U.S. Fire Administration and the National Fire Academy. As a result, fire data collection, analysis, and training were established as national concerns. Improvement in these areas will continue to occur as long as federal funding continues to support the activities.

Recommendations of *America Burning* were again revisited at a 1987 conference of fire service leaders called together by the U.S. Fire Administration. The purpose here was to review the progress made toward accomplishing the original *America Burning* report recommendations. Results showed some progress had been made toward the goals established in *America Burning*, but additional areas needed further improvement, including fire prevention, public education, and efforts to change the public's social acceptance of fire thought. The conference produced the 1987 report, ***America Burning Revisited.*** The report encouraged the nation and the fire leadership to stay on track and continue working to improve the fire service.

America Burning Revisited
The 1987 revisit to the America Burning report to evaluate the progress on accomplishing the recommended improvements.

In December 2000, the Federal Emergency Management Agency (FEMA) and the U.S. Fire Administration sponsored another conference to reevaluate the fire service and review the progress of the fire fights on those areas pointed to in the 1987 report, as still needing attention. The conference resulted in the 2000 report, ***America Burning Recommissioned, America at Risk.*** This report identified two major areas that needed correction with respect to continuing U.S. fire problems. First, it pointed to the "frequency and severity of fires or the Nation's failure to adequately apply and fund known loss reduction strategies." The report continued by recommending to the fire service that it should apply more funding and resources to fire prevention.

America Burning Recommissioned, America at Risk
The 2000 revisit of both the original America Burning report (1973) and the America Burning Revisited report (1987).

Second, the report aimed its focus on the fire service itself and pointed out that firefighters often expose themselves to unnecessary risks. The commission stated that communities "would all benefit, if the approach to avoiding loss from fires and other hazards was equal to the dedication shown on fire fighting and rescue operations." The commission recommended further education and training be conducted to assist in this transition from fire fighting to fire prevention activities (*America Burning Recommissioned*, 1987, 120–21).

Responding to the report, in the year 2000, the National Fire Academy put together a conference representing fire service educators from across the country to develop a consensus on educational requirements for the fire service. The purpose of the conference was to determine if the higher education

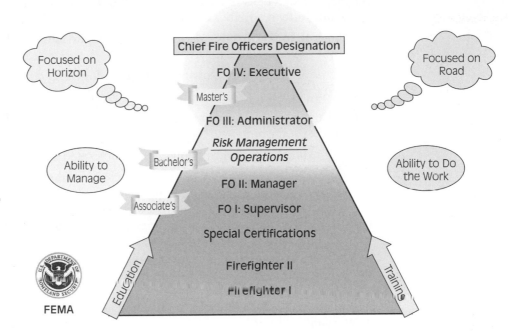

Figure 1-2 *National Professional Development Model.* Courtesy of FEMA (Federal Emergency Management Agency).

community for the fire service could develop a standardized education program for firefighters nationwide. The result was a model developed after four years of intensive discussions (see Figure 1-2). The model depicts the requirements for various fire department positions and a proposed framework for bringing together the application of training and experience to the educational system.

The adoption of a nationwide approved educational and training track for the fire service of today and tomorrow is regarded as an important step in bringing the fire service a higher degree of professionalism. The challenges of the twenty-first century can only be met by well-educated, highly trained firefighters.

Comparing the U.S. Fire Problem

After its formation in the late 1970s, the U.S. Fire Administration started focusing on the fire service by funding an investigative study to further define the extent of the fire problem in the United States. The 1982 report, *International Concepts in Fire Protection: Ideas that Could Improve U.S. Fire Safety,* examined the performance of the fire service and then compared fire loss statistics with other industrialized countries of the world. The results showed that the United States had five major differences in comparison to other highly industrialized countries that may contribute to a higher fire loss.

1. Less than 3% to 5% of the total fire department budget is spent on fire prevention—related activities.
2. Fire services receive low or inadequate funding.
3. Wood is extensively used in construction.
4. The use of plastics has increased, thus increasing the heat output of a fire over wood products.
5. U.S. social acceptance allows an uncontrolled fire to occur.

As noted first in the previous list, U.S. fire departments spend, on average, only 3% to 5% of the total fire department budget for fire prevention work. Most other countries spend an average of 5% to 15% of their budget on fire prevention work. As a result, the study recommends that more funding is needed for fire prevention.

Second, fire has traditionally been a local government responsibility, funded only by local monies which in many cases are inadequate. More recently, the federal government has approved some federal grant initiatives, but the funding is cyclical and often the result of a great deal of lobbying and pleading for federal support. Consequently, a coordinated effort to plan and develop a consistent method for national financial support for the fire services across the country has not been very promising.

British thermal unit (BTU)
A standardized measure of heat, which is the heat energy required to raise the temperature of one pound of water one degree Fahrenheit.

Third, building construction in the United Stares differs from many other countries. Because the United States has an abundance of timber, there is a greater emphasis on wood construction than masonry products. This has resulted in a large number of wooden buildings, which have little or no fire resistance rating. In comparison to other countries, wood construction has created a more serious fire problem for the U.S. fire service.

Fourth, since the early 1960s, the emphasis on the use of plastics both in furniture and building contents has led to the increase in the average amount of **British thermal units (BTUs)** per square foot of floor space. When a fire involving plastics occurs in a confined space, the opportunity of a **flashover fire** to occur is greater. Also, the amount of heat generated from plastics leads to fires that quickly consume all combustible materials in an enclosed space (see Chapter 3).

Flashover fire
A sudden event that occurs when all the contents of a room or enclosed compartment reach their ignition temperature almost simultaneously, producing an explosive fire.

This phenomenon has been complicated by the increase in the requirements for room heat conservation, which has resulted in double-paned glass with better heat retention properties for buildings. The end result in the event of a fire is quicker heat buildup in the enclosed room with little oxygen, resulting in a shorter period of time to the point of flashover.

Fifth, a fire event in the United States is a socially accepted occurrence. Fire insurance and charitable groups are available for those suffering from a loss by fire. This is in contrast to many other countries, especially those in Asia, where having a fire is considered a social disgrace. This cultural aspect brings a great deal of peer pressure from members of the community, thus persuading all community members to be more fire conscious to avoid suffering from the social disgrace associated with its occurrence.

Emergency Medical Services

There is a link between the fire service and providing emergency medical care. In this country, the fire service started in the 1930s providing first-aid units for on-scene injured firefighters. Soon, this service was extended to the public, providing first aid and assistance for those experiencing medical problems and injuries. This service was expanded to persons trapped in vehicles as the result of an automobile accident and has continued to grow and expand exponentially.

Radio communication systems improved greatly so two-way communication with hospitals and fire-dispatching centers became possible. Firefighters trained as paramedics could communicate directly with the hospital physician and, working under the physician's direction, could provide life-saving medical interventions. The fire service became an important service to the community as it expanded its available services to include emergency medical care and transportation. Today, over 70% of the requests for fire department assistance are for medical emergencies.

The emergency medical services provided by fire departments have allowed them to grow in professionalism and to meet changing community needs. In the future, further technological improvements in communication methods and scientific advancements in medicine will further increase the opportunities for the fire service to enhance the services they now provide.

Building and Fire Code Enforcement and Improvements

Even though most of the corruption in the creation and enforcement of building codes during the Industrial Revolution has been eliminated and significant improvements in building and fire codes have been made, improvements are still needed. The fire and safety report issued after the World Trade Center attack in 2001 found a number of needed improvements which would have better protected the buildings from collapse. The ensuing fires and building damage caused by the two jet airliners that were flown into the buildings brought down both high-rise buildings, killing over 3,000 people. Prior to this incident, it was thought that most fire and safety problems in modern high-rise buildings had been solved. The 2003 World Trade Center study indicates there are still issues to be addressed, such as:

- reducing the use of impact-resistant enclosures around egress systems;
- examining the steel floor, to ensure that truss bolting materials used to attach the floors to the walls are robust;
- grouping emergency egress stairways in the central building core as opposed to dispersing them throughout the structure; and
- providing resistance of passive fire protection to blast and impact in buildings designed to provide resistance to such hazards.

This report and its recommendations are important for firefighters to understand, especially the interrelationships between building construction methods, fire-resistant materials, fire combustion processes, the behavior of fire, and the impact each can have on firefighter safety.

In some cases, improvements in one area of building construction lead to unanticipated problems for firefighters. For example, consider the installation of energy-efficient windows. These new windows play a role in the high heat conditions found in fires today, because they are designed to keep heat and cool air inside the structure and keep unwanted temperatures outside the building. These windows do not break easily from high heat inside the structure, resulting in a quicker time to flashover. Firefighters attempting to break the windows for ventilation find the plastic glass materials much more difficult to break.

The expanded use of plastics has also resulted in additional heat and smoke generation, making fire fighting more dangerous. Plastics generate almost twice the amount of BTUs as wood products during the combustion process. Today, we find 40% to 50% more household furnishings are made of plastics as compared to the 1950s. Some plastics also generate more dense smoke (containing more carbon) than wood during combustion; this combination generates black or very dark smoke products. Although firefighters are now better equipped to handle these types of fires, they are still dangerous and increasing numbers of firefighters die fighting such fires each year. Understanding fire behavior and the combustion process is the first step in learning how to control and extinguish these fires.

Training and Education

Firefighters are better trained today than ever before because of improved procedures, requirements, and techniques. During the last thirty years, the fire service has required a higher level of training. For example, training firefighters for dealing with hazardous materials has become a critical and basic function in the fire service as more exotic and hazardous materials are finding their way into everyday lives.

Not only is better training for firefighters required in fire service, but also the training requirements are strictly enforced. Administrative reporting and accountability systems designed to hold those responsible for training activities have been greatly improved. Computers now track the training of each firefighter, and accountability has been improved through certification processes designed to test the results of the training program. Higher levels of education are now being recognized as a requirement for everyday fire fighting activities, and higher education will soon be required for promotion into fire service decision-making positions. When not responding to an emergency, firefighters are busy training and preparing for the next incident. In the future, more training, education, and experience will be needed to face the challenges.

Administration and Coordination

Incident Command System (ICS)
A management system utilized on the emergency scene that is designed to keep order and follow a sequence of set guidelines.

Wildland/urban interface
The line, area, or zone where structures and other human developments intermingle with undeveloped wildland or vegetative fuels.

National Incident Management System (NIMS)
A comprehensive management system designed to integrate resources from a number of public and private agencies.

A great deal of time, effort, and commitment of resources was put into the development of the **Incident Command System (ICS).** In the late 1960s, California experienced a number of **wildland/urban interface** fires that destroyed billions of dollars in residential property located between the forest/wildland area and the coastline. Since 2004, the ICS is also known as the **National Incident Management System (NIMS),** which is a more inclusive "all-risk" system (see Chapter 10).

Although there were many smaller issues surrounding the fire service's response to these fires, one of the major problems plaguing the fire service was the inability to consolidate the responding agencies' resources into one effective force to deal with the wildland fire. Some of these wildland fires would spread rapidly through four to six governmental jurisdictions, in some cases even before all the resources could be brought together to face the emergency.

These types of fires produced criticism from the press and prompted the fire agencies to join together to find a solution to these costly and sometimes deadly fires. In 1968, the U.S. Forest Service funded a grant for research and development in an effort to find a system to solve these problems. The $5 million grant was contracted out to the Rand Corporation, a West Coast think tank. Under the contract, the Rand Corporation would provide the specialized information while a committee of representatives from the agencies in Southern California would provide technical and political inputs.

The ICS, as part of the National Incident Management System, is relevant to the operations of all types of emergency incidents. It is highly recommended that readers review and become familiar with the functions and operations of this system because it is used in every emergency operation. The accompanying textbox provides a brief description of the ICS and an overview of the system's key elements.

Incident Command System: Purpose and Key Components

The purpose of the ICS is to provide for a systematic development of a complete, functional command organization designed to allow for single- or multiple-agency use. This increases the effectiveness of the command and control of an incident and provides an increased level of firefighter safety. To accomplish these important goals, the system contains the following key components.

- The systematic development of a complete, functional organization will occur, with the major functions being command, operations, planning, logistics, and finance/administration.

(continued)

- The system will be designed to allow for multiple-agency adoption by federal, state, and local fire agencies; therefore, the organizational terminology used in the system must be designed to be acceptable and workable within all levels of government.

- The system must be designed to be used as the everyday operating system of all incidents. The continuous operation permits the transition from a small incident to a large- or multiple-agency incident with a minimum of modification for participating agencies.

- The system must allow the organization for handling large fires to be built from the ground up. In addition, the start-up or beginning phases must be manageable with a minimum number of persons, with functional units designed to handle the most important incident activities. As the incident grows in size and/or complexity, unit management is assigned to allow individuals to maintain a reasonable span of control over functional as well as geographical areas.

- The system must be designed with the understanding that the jurisdictional authority of the agencies will not be delegated to other agencies. Each agency with responsibility will have full command authority within its jurisdiction at all times. Assisting agencies will normally function under the direction of the incident commander appointed by the jurisdiction within which the incident occurs or if other arrangements are agreed to by the participating agencies.

- Multiple-jurisdictional incidents will normally be managed under a unified command management organization including a single incident command post and a single incident action plan developed and applied to all participating agencies.

- The intent is to have the system staffed and operated by qualified members of any agency, and a typical incident could involve the use of personnel from a variety of agencies each working in diverse locations with the system.

- The organization can expand or contract based on the needs of the incident. Both size and complexity are used as criteria for the size of the management system.

These systems built by and for the fire service are now accepted as the operating system for the majority of the fire service as well as many other government agencies. Its continued use and refinement will prepare the fire service and other governmental agencies for emergencies and disasters well into this century.

Source: National Interagency Incident Management System (NIMS), Incident command system; Stillwater, OK. Fire Protection Publications, Oklahoma State University, 1983.

Equipment and Personnel Protection

Fire equipment has improved significantly over the past few years with the addition of powerful diesel engines, better braking systems, and larger capacity pumps. In addition, new larger-sized hose, better designed nozzles, and safer protective equipment and clothing have all made the fire service more efficient and safer for the firefighters.

Larger and lighter fire hose made with synthetic materials is one example of how technology has assisted the fire services in providing more service with less human resources. Almost all small attack lines today are 1.75 inches (45 mm) in diameter, allowing delivery of more water than the older 1.5-inch (38-mm) cotton lines could deliver. This size allows a quicker attack as the hose is lighter and more maneuverable than the 1.5-inch (38-mm) cotton hose lines.

Nozzles also have a better design to provide more water to the origin of the fire with less damage. They are more precisely controlled and tend to leak less than earlier types.

These are just a few of the many technological advances that have had an impact on the fire service since the Industrial Revolution. Perhaps one of the most significant issues currently facing the fire service is the incorporation of technology. We see vast differences in our everyday lives, and can expect that technological advances will change the face of the fire service both in the short term and in the future.

Protective Systems

Smoke and carbon monoxide detectors have been greatly improved by increasing reliability and reduced costs through mass production. The lower costs have made them available to greater numbers of citizens. These warning devices have been responsible for a significant reduction in fire deaths by providing an early warning of dangerous conditions, thus allowing people time to escape the structures where they are installed.

The development of the quick-acting fire sprinkler head along with new and improved residential sprinkler systems continues to impact residential fire losses.

The increased use of protective systems in residential, commercial, and multifamily dwellings is another key component to reducing deaths and injuries from fires. Their introduction into the mainstream has caused a change in response by fire departments, and will likely continue to do so in the future.

FIRE SERVICE OF THE FUTURE

The twenty-first century offers many challenges as well as exciting new technologies to the fire service. To move forward, the U.S. fire service needs to complete a long past due commitment to the adoption and use of the International System of Units (SI) measuring system.

In the 1970s, it was generally agreed that the measuring systems used in the United States would be converted from the English system to the SI system to be consistent with most fire departments that are already using this system. As worldwide communication systems continue to improve and expand, the fire service in countries around the world will be able to further improve communications.

For the U.S. fire service of this century to stay in tune with fire departments around the world, it must begin to use the SI system in daily routines. The SI system will slowly become the system of main use and the fire service will be able to effectively communicate with those already using the system. A brief introduction of the SI system and a comparison chart are provided in the following textbox.

SI Measuring System

Firefighters are required to use numbers in the performance of their work. For example, a numerical system is used to describe the size of hose lines employed in fire attack. For house fires, 1.5-inch (38-mm) or 1.75-inch (45-mm) lines are used. A single 2.5-inch (64-mm) hose line will be used to supply water, while the capacity of the engine pump is described as a 1,500 gpm (5,678 l/min) centrifugal pump.

For these numbers to make sense, they must be used with some unit of measurement that describes what is being measured. To do so, we use three main units of measure: distance, mass, and time.

Distance, Mass, and Time

In the United States, the fire service is still using the English or customary system to measure. Most other nations and the scientific community use a form of the metric system called the International System of Units or SI (after the French, *Système International d'Unitès*). This method was developed early in the nineteenth century in an attempt to standardize the way length, distance, weight, and volume are measured. It is commonly used in most nations of the world; however, the United States continues to use the English system. It is imperative that the U.S. fire service changes to the SI system of measurement to better exchange information and communicate effectively with fire organizations worldwide.

The basic unit for length in SI is the meter (m). It is also the basic unit of length in the metric system and is almost equivalent to the English yard. Whereas the meter is 39.7 inches in length, the English yard is 36 inches. From this base unit, area can be derived in square meters (m^2) and volume in cubic meters (m^3).

(continued)

Measurements for speed can be derived from length and time and described in meters per second (fps or m/s). The other unit of concern is the measurement of length. In the customary system, the inch is one-twelfth of a foot, the yard is 36 inches or 3 feet, and the mile is 5,280 feet or 1,760 yards.

Because all units of surface and volume capacity are developed using the meter standard, the relationship between all units is understandable as they are both divided and multiplied by factors of 10. In the metric system, cube subdivision is used to describe the relationship of length, area, and volume. The cube is one decimeter square and contains 1,000 cubic centimeters.

Energy and Work

Energy has been defined as the capacity to perform work. Work occurs when a process is applied to an object over a distance. In the SI system, the unit for work is the joule (J). The joule is derived from the unit of $kg\,m/s^2$ and the distance in meters.

In the English system, the unit of work is the foot-pound (ft lb).

Fire Service and Technology

In 2003, the U.S. Fire Administration was funded by legislation "to provide for the establishment of scientific basis for new fire fighting technologies standards; improve coordination among Federal, State and local fire officials in training for and responding to terrorist attacks and other national emergencies and for other purposes." These funding and other research dollars are now being provided to improve the technologies for fire service use. More dollars and research can be expected in the near future.

To provide direction for these expenditures, an emergency response technology (ERT) advisory group was established by a FEMA grant to "identify the needs of emergency responders" and then search for solutions for those needs. The advisory group works with federal laboratories, universities, and private industry on research and technology projects. Some of these emerging technologies and research efforts are briefly discussed below.

Alternative fuels

Presently two types of compressed gases are being used for motor vehicle fuel.

Alternative Fuels The use of **alternative fuels** for vehicles has been increasing in the United States as part of an effort to reduce air pollution and dependence on foreign oil. Compressed natural gas (CNG), liquefied natural gas (LNG), and liquefied petroleum (LP-Gas) are currently the most commonly used alternative fuels in the United States. Research into hydrogen-fueled vehicles is just getting underway and as the technology improves, the use of hydrogen as a fuel for vehicles will certainly be more widespread.

Firefighters need to prepare for fire fighting activities at alterative fuel storage/dispensing locations as well as fires in vehicles powered by these fuels of the future. Understanding the nature of fire chemistry and the fire combustion processes will greatly improve firefighters' opportunities to extinguish fires in these materials (see Chapter 9).

Infrared imaging
A method using infrared waves to detect heat being radiated by a substance or body.

Infrared Imaging   Today **infrared imaging** with portable cameras is proving to be a useful technology which can be applied by the fire service to assist in discovering the location of trapped or downed persons, a fire working its way through a building, a deep-seated fire in a grain storage silo or coal storage bin, or the active edge of a wildland fire.

Because the human body gives off heat, the infrared camera can also be used to seek out and locate persons who are lost, trapped, or somehow incapacitated in building fires. They have also proven their importance in urban rescue operations where the collapse of the building has trapped persons under mounds of debris. Using a probe to penetrate the materials, the heat-seeking device can be positioned into the debris to detect heat given off by the body. Infrared cameras can also provide a warning of impending flashover, but to read these signs, firefighters must be skillful in the use of the camera.

Today most infrared imaging cameras can penetrate only lightweight construction materials, so there are limitations on their ability to locate a fire or person inside some structures. However, under new federal legislation, the U.S. military has been authorized to release the latest infrared technology to local governments. This newer technology provides greater penetrating ability so fires and persons will be able to be found in heavier construction materials. This is just one small example of why future firefighters need to be better trained and educated in order to understand and operate this type of high-technology equipment.

Global positioning
Using satellites positioned in outer space, locations on earth can be triangulated by the beam from several satellites.

Global Positioning   The use of **global positioning** systems will become more prevalent in the fire service as the technology continues to improve. This system will enable fire managers and communications centers to determine the fire equipment closest to an emergency and then direct that company to the fire. Connected with a computer-aided dispatching system, the tracking and availability of the companies nearest to the incident can be dispatched as additional companies are needed. This tracking ability is very important as fire companies are being pressed into emergency medical service and their availability for fires has been significantly reduced. A quick tracking and availability system provides real-time information and allows better coverage with fewer units because their location and availability can be tracked and monitored.

Global positioning systems will assist a preprogrammed computer system to apply mapping systems for fire modeling. For wildland fires, the computer can be preprogrammed with fuel, topography, and weather and fire history

information for projection of fire behavior in real time or time intervals on a map of the area to assist the incident commander in planning and decision making.

Ultrafine Water Mist The banning of Halon 1301, a halogenated-based extinguishing agent developed and put in service in the 1960s and 1970s, was later identified as an ozone-depleting substance. This resulted in the search for an alternative extinguishing agent. It seems that while some non-ozone-depleting agents have been discovered, they too have limitations that may result in serious aftereffects. From an environmental and toxicity viewpoint, ultra mist-based water systems seem to be a good alternative as an extinguishing method.

Currently, the use of water in an **ultrafine water mist** is being perfected. The water mist, or application of water at very high pressure, is accomplished through fixtures specifically designed to finely separate the water into very small particles in order to improve the conversion to steam. Also, the gaslike dispersion behavior adds to its ability to act as a total flooding agent.

The efficiency of water mist fire suppression depends critically on mist size, mist stability, the transport behavior of mist in an obstructed space, and efficiency and rate of droplet vaporization.

Although not available for portable use at this time, this high-pressure application of water may be perfected for future use by firefighters. It will require further refinements in nozzles, firefighter clothing protection (from steam burns), and high-pressure pumps for the apparatus. High-pressure pumps were supplied on engines during the 1960s, but the pump reliability and nozzle technology were not perfected to a point where firefighters could consistently depend upon the system. They were phased out during the 1970s and 1980s.

Compressed Air Foam Systems The application of Class A foam using air under pressure is a developing technology that is designed to assist in the application of Class A foam at the scene of structure fires, wildland fires, and **Class B fires.**

The **compressed air foam system (CAFS)** is a Class A hydrocarbon-based **surfactant** that improves the wetting and penetrating capabilities of plain water because it reduces the water's surface tension. In addition, it consists of water and additives formed around air bubbles, making it a durable wet but light extinguishing agent. Because of this characteristic, less water is required to extinguish fires. It makes hose lines lighter and more moveable, and reduces the amount of water used and resulting water damage. Engines are now being equipped with systems using these foaming materials; further improvements in this technology will result in a wider usage and application.

Improvements in Building Safety

Research is now being conducted to support current and long-term improvements to reduce the vulnerability of the structure, building occupants, and first

---

Ultrafine water mist
Water is dispensed under high pressure through very fine nozzle outlets creating nearly micro-sized droplets.

Class B fires
Fires involving flammable liquids.

Compressed air foam system (CAFS)
A developing technology that includes a Class A hydrocarbon-based surfactant that improves the wetting and penetrating capabilities of plain water.

Surfactant
A soap material that works to ease the surface tension of water, allowing the water to more easily penetrate materials and thus increasing the effectiveness of water as an extinguishing agent.

responders to extreme threats. Following are only some of the areas that are targeted to research in further depth.

1. **Increasing structural integrity.** Structural integrity can be increased through the development and implementation of performance criteria for codes and standards, tools, and practical guidance for prevention of progressive structural collapse. The collapse of the World Trade Center towers has triggered a new look at the entire subject of structural safety.

2. **Enhancing fire resistance.** Fire-resistant steel is available and in use in other countries today. More efficient and accurate testing methods for performance of steel under building fire conditions are needed in the United States to support the incorporation of fire-resistant steel into U.S. construction practice.

3. **Improving emergency egress and access.** After the events of 9/11, it was realized that current egress models may be inappropriate and/or insufficient for the design and placement of doors and stairways and the control of elevator movement. In the future, behavioral and engineering studies will be conducted to enable the development of simulation tools to better capture the movement of people within a building under fire and other emergency situations.

4. **Cybernetic building systems.** In the future, improvements in technology and increasing economic costs of building construction will encourage companies developing building controls to improve performance and reduce their costs.

Cybernetic building systems
Automatic control systems installed in buildings to integrate the building services in a central location including energy management (HVAC systems), fire detection and security systems, building transportation systems, fault detection, and diagnostics.

These cost reductions can be achieved by developing **cybernetic building systems** which integrate more building services, including energy management, fire and security, transportation, fault detection, and real-time diagnostics of the building. The ability to assess and reduce the vulnerability of buildings using indoor air quality (IAQ) assessment simulations to detect chemical, biological, and radiological aerosols and other dispersion methods inside the structure will also be included.

For the fire service, these systems will provide on-site application of fire systems within the building as well as faster fire department notification. As the sophistication of these systems increases, so will the need for fire service personnel to be educated and trained to understand and respond to these complicated systems.

It is incumbent upon the fire service to start preparing itself for this higher level of technology to be ready for the challenges of this century.

SUMMARY

The United States has a long history of conflagrations and large fires. The lack of social pressure, an abundance of timber for construction, the rapid growth of deficiently planned cities, and poor building code and fire code enforcement all contributed to the frequency of these events.

The fire service has since expanded the application of new technology, improved the training and education of firefighters, improved fire fighting equipment, and developed an Incident Command System to bring together diverse emergency resources when needed.

As we move through this century, continued improvement is vital. The U.S. fire service needs to embrace the SI measuring system. It needs to continue enhancing the fire resistance of building construction materials and explore further uses of technology.

It is important to look at fire service history and learn from mistakes made. An understanding of modern fire service needs and additional knowledge of the fire combustion processes are required. Also, we must realize that the future of fire service requires a vision that includes a higher level of training, education, and the embracement of technological advances.

KEY TERMS

Alternative fuels   Presently two types of compressed gases are being used for motor vehicle fuel. One is compressed liquefied natural gas (CNG) and the other is compressed liquefied petroleum gas (CLP) (either propane or butane). Research is now being conducted to expand the use of hydrogen gas. These gases are under high pressure and present a danger of fire and/or explosion when released.

America Burning   The 1973 report from a committee appointed by the president to investigate and report back its findings regarding U.S. fire service. Over ninety recommendations were forwarded for action.

America Burning Recommissioned, America at Risk   The 2000 revisit of both the original *America Burning* report (1973) and the *America Burning Revisited* report (1987). Two areas noted as needing further attention were fire prevention and firefighter safety.

America Burning Revisited   The 1987 revisit to the *America Burning* report to evaluate the progress on accomplishing the recommended improvements.

British thermal unit (BTU)   A standardized measure of heat, which is the heat energy required to raise the temperature of one pound of water one degree Fahrenheit.

Class B fires   Fires involving flammable liquids.

Compressed air foam system (CAFS)   A system that produces high-quality foam bubbles which are small and consistent in size and density. The system uses an air compressor to provide air at a uniform pressure and an automatic foam proportioner, which ensures the foam solution will be kept at the correct ratio of water to foam solution as set by the operator. This automatic control over the mixture allows the operator to control the consistency of the foam generated.

Conflagration   A fire with major building-to-building flame spread over a great distance.

Cybernetic building systems   Automatic control systems installed in buildings to integrate the building services in a central location including energy management (HVAC systems), fire detection and security systems, building transportation systems, fault detection, and diagnostics.

Flashover fire A sudden event that occurs when all the contents of a room or enclosed compartment reach their ignition temperature almost simultaneously, producing an explosive fire.

Global positioning Using satellites positioned in outer space, locations on earth can be triangulated by the beam from several satellites. This triangulation allows accurate identification of a position on the earth to be determined. Fire mapping systems can be integrated into the system for computer modeling on wildland fires and for tracking engine companies available for response.

Incident Command System (ICS) A management system utilized on the emergency scene that is designed to keep order and follow a sequence of set guidelines.

Infrared imaging A method using infrared waves to detect heat being radiated by a substance or body. Firefighters use a camera to translate the location and intensity of the heat being given off. Using these cameras they can locate victims or other firefighters who may be trapped.

Insurance Services Office (ISO) An agency funded by the insurance companies to independently apply the grading schedule to cities (fire departments) and set the rate for fire insurance premiums.

National Incident Management System (NIMS) A comprehensive management system designed to integrate resources from a number of public and private agencies. It encompasses most of the Incident Command System as one component of the overall management system.

Surfactant A soap material that works to ease the surface tension of water, allowing the water to more easily penetrate materials and thus increasing the effectiveness of water as an extinguishing agent.

Ultrafine water mist Water is dispensed under high pressure through very fine nozzle outlets creating nearly micro-sized droplets. This fine mist exhibits high-energy absorption behavior because of the high vaporization rate of the fine water particles (the surface area of the droplet exposed to the heat). Additionally, the gaslike dispersion of the mist acts like a total flooding agent as the fine mist is dispersed throughout the enclosure.

Wildland/urban interface The line, area, or zone where structures and other human developments intermingle with undeveloped wildland or vegetative fuels.

REVIEW QUESTIONS

1. List the four main public concerns identified during "the decade of conflagrations" that are still applicable to today's fire service.

2. Identify the five major differences between the United States and other industrialized countries that contribute to higher fire losses.

3. Describe the major differences between the Incident Command System (ICS) and the National Incident Management System (NIMS).

4. Describe how the events of 9/11 help identify improvements to building and fire codes.

5. How does a global positioning system help in the fire service?

REFERENCES

Angle, James et al. (2008). *Firefighting Strategies and Tactics* (2d ed.) (Clifton Park, NY: Delmar Cengage Learning).

Delmar Cengage Learning. (2008). *The Firefighter's Handbook: Firefighting and Emergency Response* (3d ed.) (Clifton Park, NY: Delmar Cengage Learning).

Gagnon, Robert M. (2008). *Design of Special Hazard and Fire Alarm Systems* (2d ed.) (Clifton Park, NY: Delmar Cengage Learning).

Insurance Services Office. (1998). *Fire Suppression Rating Schedule* (Chicago: ISO).

Manning, Bill (ed.). (1997, February). "Around the Fire Service, 1879–1889," *Fire Engineering*, 57–61.

———. (1997, June). "Around the Fire Service, 1909–1919," *Fire Engineering*, 74–79.

———. (1997, July). "Around the Fire Service, 1930–1939," *Fire Engineering*, 79–84.

———. (1997, September). "Around the Fire Service, 1950–1959," *Fire Engineering*.

———. (2000, September). "History of the Fire Service, The Last 120 Years," *Fire Engineering*.

National Institute for Standards and Technology, http://www.fire.nist.gov (regarding cybernetic building systems).

Schaenman, Philip. (1982). *International Concepts in Fire Protection: Ideas that Could Improve U.S. Fire Safety* (Arlington, VA: Tri-Data Corporation).

———. (1993). *International Concepts in Fire Prevention* (Arlington, VA: Tri-Data Corporation).

———. (1999). *Profile of the Urban Fire Problem in the United States* (Arlington, VA: Tri-Data Corporation).

Society of Fire Protection Engineers. *Fire Protection Engineering,* http://www.pentoncmg.com/sfpe/index.html

U.S. Fire Administration

The following reports as well as many others are available to members of the fire service at *https://www.usfa.dhs.gov/applications/publications/*.

FEMA. (1973). *America Burning: The Report of the National Commission on Fire Prevention and Control* (Washington, DC: FEMA).

———. (1987). *America Burning Revisited* (Washington, DC: FEMA).

———. (2000). *America Recommissioned, America at Risk: Findings and Recommendations on the Role of the Fire Service in the Prevention and Control of Risks in America* (Washington, DC: FEMA).

Chapter 2

FIRE CHEMISTRY

Learning Objectives

Upon completion of this chapter, you should be able to:

- Understand and explain the basic structure of atoms.
- Explain how atomic structure determines the behavior of elements and compounds.
- Understand basic chemical and physical properties and concepts and how they influence the behavior of materials involved in fires and hazardous materials incidents.
- Correlate chemical structure with chemical names to allow for a general prediction of some hazardous chemical behaviors.
- Understand key physical properties of chemicals and how these properties are related to fire protection.

INTRODUCTION

This chapter reviews basic chemistry and physical processes that impact the physical and chemical properties of materials. Firefighters are faced with an increasing use of hazardous materials and, as a result, more opportunities are being presented for leaks, spills, and incidents involving these materials. It is essential for the safety of firefighters to have a basic understanding of these chemicals and their reactions during the fire combustion process.

There are numerous examples of firefighters unknowingly encountering a chemical reaction which resulted in the sudden violent release of energy or toxic vapors. Such incidents could endanger the lives of those in the community as well as first responders. This chapter prepares first responders by providing definitions, basic concepts, and descriptions of the physical and chemical processes of fires and chemicals that may be encountered.

MATTER

Matter
Anything that occupies space and has mass or something that occupies space and can be perceived by one or more senses.

Matter is anything that occupies space and has mass or something that occupies space and can be perceived by one or more of the senses. It also can be best described by its physical appearance or by its physical properties that can be measured and observed such as mass, size, or volume. Firefighters observe matter generally in three states: solid, liquid, or gas. In addition, firefighters can detect it in some cases by color or smell. When matter is spoken of as having **mass,** it is said that mass is a property of physical objects. It is basically a measure of the amount of matter the objects contain. In informal usage, the word *weight* is often used synonymously with *mass*.

Mass
A measurement of quantity when the weight is proportional to the mass.

States of Matter

In defining matter, it is usual to classify it into three divisions or states of appearance: solids, liquids, and gases. In these states, the material can act and appear very differently depending on its present state.

- **Solids.** A solid consists of a portion of matter, which has a definite volume and a definite shape.
- **Liquids.** Liquid matter has a definite volume but not a definite shape. A liquid will take on the shape of the vessel in which it is contained. Some liquids will generally turn to a gas when exposed to the atmosphere and others can be turned into a gas when placed under pressure and heat.
- **Gases.** Matter, or *gas,* has neither a definite shape nor volume. Its shape is determined by the amount of pressure placed upon the gas and the shape of the container. The laws pertaining to gases are covered later in

this chapter, including Boyle's and Charles's laws. Flammable gases are of special concern to firefighters because they may be encountered on emergency responses.

A **flammable gas** is one that is flammable at atmospheric temperature and pressure within a mixture with air of 13% or less (by amount or volume) or it has a **flammable range** with air of more than 12% (ERG 2008).

Chemical Properties of Matter

All materials are classified as either **inorganic** or **organic.** An inorganic classification means that the matter is comprised chiefly of earth minerals such as rocks, soil, air, water, and minerals in or below the earth's surface. Examples of inorganic substances are quartz, sulfur, iron, and granite. They are referred to as minerals. A majority of inorganic minerals are not involved in the combustion process.

An organic classification generally means that the matter is found in substances that were once living organisms. Organic substances consist of carbon, hydrogen, and oxygen. Examples of organic substances are plastics, wood, gasoline, and oil. Matter that was organic or a living organism at one time contained cells used to grow and feed the organism.

Cells

Cells are a tiny mass of protoplasm that usually contains a nucleus, which is enclosed by a membrane and forms the smallest structural unit of living matter. This structure is capable of functioning independently. In humans, cells make up the organs of the body. The organs then are interconnected to the entire body through ten systems. The subject of organic chemistry is important to firefighters not only for understanding fire behavior but also for understanding the functioning of the human body.

COMPOUNDS

In chemistry, a **compound** is a substance formed from two or more elements joined with a fixed ratio. In general, the ratio must be fixed due to physical property. The defining characteristic of a compound is its chemical formula. Formulas describe the number of atoms in a substance.

A compound consists of molecules, which are combined chemically. They are homogeneous and have a definite composition regardless of origin, location, size, or shape. The elements in a compound cannot be separated by any physical means. Compounds are also more abundant than elements, as there are approximately 5 million compounds that have been identified. An example of a compound is

---

Flammable gas
A gas that is flammable at atmospheric temperature.

Flammable range
The numerical difference between a flammable substance's lower and upper explosive limits in air.

Inorganic
A classification in which the matter consists primarily of the materials of the earth.

Organic
A classification in which the matter is found in living substances such as plants and animal life.

Cell(s)
The objects that comprise all organisms, numbering one or more, where all vital functions of the organism occur within the walls.

Compound
A substance formed from two or more elements joined with a fixed ratio.

wood, a complex material that is mainly composed of cellulose, which consists of carbon, hydrogen, and oxygen with lesser percentages of nitrogen and other elements. This combination of atoms is generally agreed to be six carbon atoms, ten hydrogen atoms, and five oxygen atoms, represented by $C_5H_{10}O_5$. This combination can vary slightly depending on the amount of oil and the density of the wood.

All compounds will break up into smaller compounds or individual atoms with the application of heat. The temperature the material must be heated to in order for it to decompose is termed the *decomposition temperature*.

Structure of Molecules

Molecule
Two or more atoms tightly bound together by chemical bonds.

Molecules are made up of atoms. Atoms are smaller and are joined together using a form of electricity to bond. Figure 2-1 shows a piece of wood that is subdivided into a cell, molecules, and the cell wall which contains the molecules and the compounds that make up the wood. Note the complexity and interworking relationships needed to produce wood.

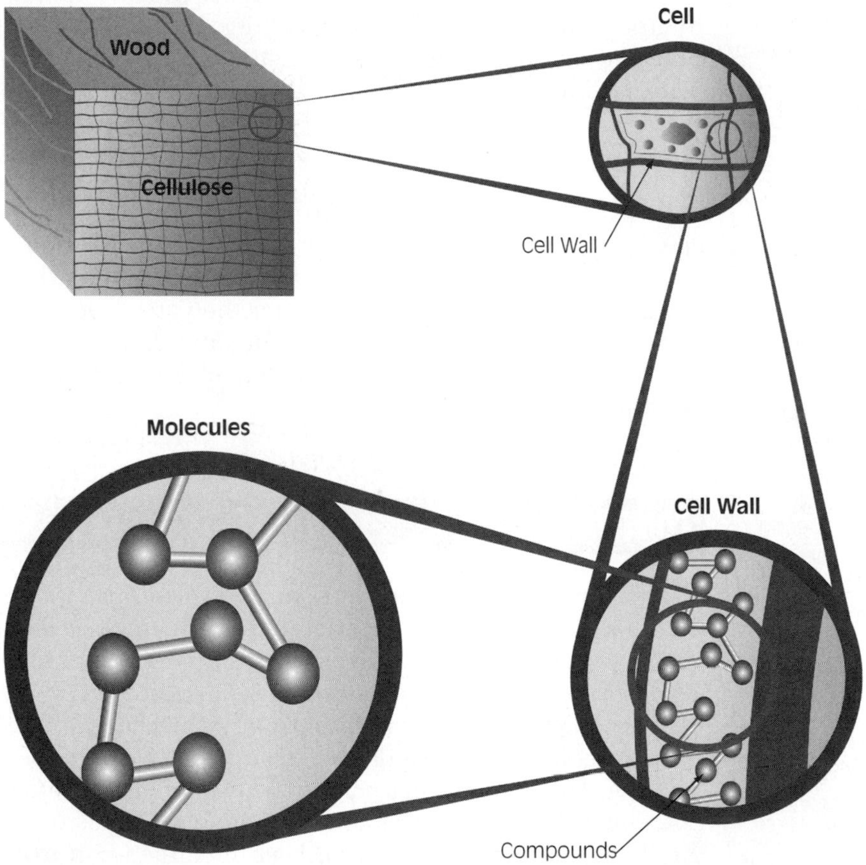

Figure 2-1 *Wood subdivided into the cell, molecules, and compounds.*

Atom
Made up of a neutron that is electrically neutral, a proton that has a positive electric charge, and a surrounding cloud of orbiting electrons, which are negatively charged.

Neutron
A particle with nearly the same mass as the proton, but electrically neutral.

Proton
A positively charged particle that is the nucleus of the most common isotope of hydrogen.

Electron
A very light particle with a negative electrical charge, a number of which surround the nucleus of most atoms.

Negative ion
Forces outside an atom cause it to gain an extra electron, giving it an overall negative charge.

Positive ion
An atom that is missing electrons.

Element
The simplest form of matter.

ATOMS

Atoms are the basic buildings block of matter. They are the smallest unit of an element that takes part in a chemical reaction. Atoms of nearly every element can combine with other atoms to form molecules by chemical interaction, whereas a molecule is the smallest unit of an element or compound that retains the chemical characteristics of the original substance. As an example, water (H_2O) is made up of the combination of two atoms of hydrogen and one atom of oxygen. They combine, creating a molecule that differs from both the original atoms (hydrogen and oxygen).

Structure of the Atom

Atoms consist of three types of subatomic particles. At the center of an atom is the nucleus, which contains two kinds of particles: **neutrons** and **protons.** The neutron is heavy and has no electrical charge. The proton is equal in weight to the neutron and contains a positive electrical charge. Outside the nucleus, spinning around it in orbit, are the **electrons.** These are light and carry a negative charge.

In a normal state, the electrical charge of the atom as a whole is zero, which means an atom contains the same number of electrons as protons. Electron orbits are arranged in layers, or shells. The inner shell is full when it contains two electrons and does not seek more. The number of electrons in the outer shell of an atom is important because a shell that is either overfull or not full seeks other electrons, and this leads the atom to react with other atoms. For example, sometimes forces outside an atom cause it to gain an extra electron. It now contains more negative electrons than protons, so the sum of the electrical charge is negative. The atom has become a **negative ion.** It is now likely to interact with another atom that is missing electrons, a **positive ion,** to return its overall charge back to zero.

Combinations of Atoms

There are ninety-two natural elements and a few more man-made elements for a total of 109 elements (Friedman, 1998). Atoms combine with other atoms in many different ways to form molecules. Atoms combine with other atoms to fill their outer electron shells, or rid themselves of extra electrons. Many large atoms have numerous electrons, as many as one hundred or more. Regardless of their size, the atom's behavior is determined primarily by the number of electrons in the outer shell.

ELEMENTS

Elements are the basic construction materials of matter. An element is the simplest form of matter. Elements can be grouped into families by their similar properties. For example, atoms of the halogen elements may be attached to various

Halogenation or halogenated
Any chemical reaction in which one or more halogen atoms are incorporated into a compound.

other atoms to form different compounds, often called **halogenated** compounds. In this grouping of compounds we find chlorine, bromine, and fluorine. These compounds have efficient and effective fire-extinguishing capabilities, but have other aspects that have resulted in the banning of their use. See Chapter 4 for further discussion of halogenated extinguishing agents.

Element Symbols

Symbols have been selected to allow us to write out chemical formulas without difficulty. Each chemical has been given a one- or two-letter abbreviation. The symbols are made up of a principle letter or letters in the name of the element. For example, H represents hydrogen and O represents oxygen.

The elements that have been discovered are identified with symbols representing the Latin names given to the element. Note the following examples.

    copper (cuprum)—Cu

    gold (aurum)—Au

    iron (ferrum)—Fe

    lead (plumbum)—Pb

    mercury (hydragyrum)—Hg

    potassium (kalium)—K

    silver (argentum)—Ag

    sodium (natrium)—Na

    tin (stannum)—Sn

Firefighters may respond to incidents where knowledge of these symbols and chemical formulas may be helpful in identifying how to deal with the situation. See the section on properties of chemicals later in this chapter for further discussion.

MOLECULES

Physical change
Molecules of a material are not changed by a process.

Chemical change
Molecules of a material are changed by a process.

As mentioned earlier in the chapter, a molecule is the smallest part of a pure chemical substance that has all the properties of the material. Molecules are in constant motion, but some are moving rapidly and some slowly, depending on the state of the material. For example, molecules in solid materials move slowly, whereas they move more rapidly in liquids. They move very rapidly in gases. The application of heat also increases this movement.

When a **physical change** occurs, such as a change from solid to liquid or liquid to gas, the molecules remain intact. When a **chemical change** occurs, the molecules are altered. These altered molecules do not have the same properties as those of the original. An example of a physical change of a fuel is the injection of diesel fuel into a cylinder cavity. The cylinder moving upward compresses the fuel into a gas and the gas explodes from the heat of compression. The pressure from the explosion forces the piston downward.

A chemical change occurs when cement, water, sand, and stone are mixed together in correct proportion. When the cement begins to heat up, undergoing a chemical change, it drives off the water which forms a new material called concrete.

One of the best and most commonly used examples of the changes that occur in the physical state of a substance is water. At normal atmospheric pressure and temperature above 32°F (0°C), water is found as a liquid. At sea level, atmospheric pressure is defined as 760 mm of mercury, which is measured by using a **barometer.**

Barometer
An instrument for measuring atmospheric pressure.

When the temperature of water falls below 32°F (0°C) and the pressure remains the same, water changes state and becomes a solid called ice. At temperatures above the boiling point 212°F (100°C) water changes state to a gas called steam.

MIXTURES

Most natural forms of matter are mixtures of pure substances. A mixture is a combination of substances held together by physical rather than chemical means. Soil and rock, plants, animals, coal, oil, air, and cooking gas are all mixtures. Mixtures differ from compounds because ingredients of a mixture retain their own properties, meaning that the substance has not been changed in the formation of the mixture. Thus, mixtures can be separated from other ingredients by physical means. An example of a mixture is a fruit salad. The combination of apples, oranges, and walnuts with dressing forms one salad. While they are combined, each of the three components still retains its own properties.

Chemical Names

Each chemical element has a unique name as an identifier. Names complying with the International Union of Pure and Applied Chemists guidelines are those responders should use when looking for information about a product. The technical names and proper shipping names can be found in the DOT 49 CFR Section 172.101, Table of Hazardous Materials.

For international trade relations, the official names of the chemical elements (both ancient and recent discoveries) are decided by the International Union of Pure and Applied Chemistry Group, which has approved the use of

(continued)

the English language. There have been a few recent changes. For example, *aluminium* and *caesium* are now to take the place of the U.S. spellings *aluminum* and *cesium,* and the U.S. spelling of *sulfur* is replaced with the British spelling *sulphur.* These changes take place on a continuing basis as the global economy moves to accommodate more countries. Firefighters need to continue updating their knowledge and information bases as well. Chapter 10 will discuss more about finding chemical names and identifying their hazards.

Prefixes and Suffixes

Functional groups
Elements of the group that have the same electronic structure in the outermost shell, thus allowing the elements to show similar chemical behavior.

Prefixes (syllables added to the beginning of a word) and suffixes (syllables added to the end of a word) are used to describe atoms that are added to a basic molecule. Suffixes often denote groups of atoms which are added and behave as a group, not as individual atoms. These groups are called **functional groups.** Firefighters need to know that prefixes and suffixes are identifiers used to provide a warning. Table 2-1 shows some common substances firefighters may encounter with suffixes and prefixes. For example, a material containing additional oxygen, ammonium perchlorate, contains four additional atoms of oxygen, which makes the substance highly unstable.

Organic Chemicals

The words *organic* and *inorganic,* although not part of chemical names, are often used to describe the makeup of compounds. Organic compounds contain some form of hydrocarbons or a combination of carbon and hydrogen molecules. The number of carbon atoms in combination with the number of hydrogen atoms will determine the properties of the substance and more importantly how it will react under varying conditions. Organic compounds may have one or many carbons

Table 2-1 *Prefix and suffix—warning signs for firefighters.*

Substance	Chemical	Description	Prefix	Suffix
Chlor**ine**	Cl	Chlorine alone		**-ine**
Sodium chlor**ide**	NaCl	Chlorine and sodium (NA)		**-ide**
Sodium chlor**ite**	$NaClO_2$	Chlorine w/sodium and two oxygen (O)		**-ite**
Sodium Chlor**ate**	$NaClO_3$	Chlorine w/sodium and three oxygen (O)		**-ate**
Ammonium **per**chlor**ate**	NH_4ClO_4	Ammonia w/perchloric acid (4 atoms of H and 4 atoms of O: a powerful explosive)	**per-**	**-ate**

Source: Fred C. Hess, ***Chemistry Made Simple*** (New York: Doubleday Publishing, 1996).

linked in the makeup of the molecule. Carbon is a remarkable element. It has a great affinity to form multiple bonds or chains of atoms that may be straight, branched, or formed into rings. These carbon atoms have six electrons in the outer ring, making it capable of forming multiple bonds. Because of these properties, carbon is known to form nearly 10 million different compounds.

Organic peroxides are important because they can react explosively when involved in the combustion process. The reason these substances pose a threat is found in their chemical structure. An oxidizing agent is a substance that gains electrons in an oxidation–reduction reaction, while the reducer loses an electron. These reactions are often called *redox* reactions. When they occur, organic peroxides are chemically bonded with an oxidizing agent or additional oxygen. The oxidizer provides the oxygen and the reducer is the substance that breaks the material down, allowing it to become further oxidized. When an oxidizing agent is heated a violent reaction may occur. For example, benzoyl peroxide in an undiluted form will ignite readily and burn rapidly, almost as rapid as black powder. Heat will cause it to decompose rapidly, and if it is confined it will explode. Decomposition can also be initiated by heavy shock or heating by friction. This type of heating occurs by rubbing the surface of the material which moves the molecules creating heat.

Firefighters need to identify facilities in their jurisdictions where these products are stored, used, or transported to or from another location. In addition, they need to use the hazardous materials warning labeling system to identify hazardous materials and then follow up with more detailed information on the specific material they may encounter (see Chapter 10).

PROPERTIES OF CHEMICALS

Because all atoms of a certain element have the same structure, they also have the same properties. The structures and properties of atoms are what make their behavior predictable, which is a valuable tool for firefighters when working with hazardous chemicals.

We can combine atoms to create a chemical compound. The defining characteristic of a compound is its chemical formula. Formulas describe the ratio of atoms in a substance and the number of atoms in a single molecule of the substance. For example, ethane is described as C_2H_4—two carbon atoms and four hydrogen atoms rather than CH_2—one carbon atom and two hydrogen atoms.

Some compounds are flammable; some are nonflammable. For example, the chemical formula for ethylene oxide, a very flammable gas, is C_2H_4O (two carbon atoms, four hydrogen atoms, and oxygen), while carbon tetrachloride CCH_4 (nonflammable) always consists of one carbon atom with four chlorines, which are attached to the carbon atom. Because a compound is always made of the same atoms in the same ratio, it always has the same properties.

Physical and chemical properties determine the behavior of elements and compounds. It is important to note the temperature scale used when assessing the properties. Some properties are very temperature dependent, which should alert firefighters to a danger from the ambient temperature at the incident location. The Fahrenheit scale is used to describe the weather and ambient temperature during fire department operations, but other countries and most scientific references use the Celsius scale. In this text both scales will be used in examples and charts to help firefighters become accustomed to the Celsius scale (see Chapter 1 and Appendix A).

Boiling Point

Boiling point (BP)
The temperature at which a liquid will convert to a gas at a vapor pressure equal to or greater than atmospheric pressure.

The **boiling point (BP)** can be defined several ways. One explanation is the temperature at which the vapor pressure of the liquid equals the pressure of the atmosphere around it. Another way to explain boiling point of a liquid chemical is the temperature at which the molecules in a liquid are heated to a point that they begin to break the surface tension of the liquid and are released to the atmosphere (it begins to bubble) and change to vapor. See Table 2-2 for a list of normal boiling points of specific liquids. Once started, and as long as the heat is applied, the liquid continues to boil and change to a vapor. Table 2-2 provides boiling points in both Celsius and Fahrenheit temperatures which firefighters are likely to encounter.

Vapor Pressure

Vapor pressure (VP)
The pressure placed on the inside of a closed container by the vapor in the space above the liquid which the container holds.

Vapor pressure (VP) is the pressure placed on the inside of a closed container by the vapor or molecules being driven off the flammable liquid in the space above the liquid. It takes heat energy to release the molecules through the surface of the liquid. This energy is in the form of heat. As temperature of the liquid is raised, the molecules begin to move rapidly and some are released at the surface and continue to be released more quickly as the temperature within the liquid increases.

Table 2-2   *Normal boiling points.*

Substance	Temperature °C/°F
Benzene	80.1°C/176.18°F
Carbon dioxide	−78.5°C/−109°F
Chlorine	−34.1°C/−29°F
Hydrogen	−252.8°C/−423.17°F
Oxygen	−183.0°C/−297.3°F
Water	100.0°C/212°F

Open cup test
Measures the release of the vapors in terms of the pressure exerted at a specific temperature.

Flash point
The temperature of a liquid which, if an ignition source is present, will ignite only the vapors being produced by the liquid creating a flash fire.

Fire point
The lowest temperature at which a liquid produces a vapor that can sustain a continuous flame.

Closed cup test
Place a lid on the cup and take the pressure reading when the vapors are released from the liquid.

Vapor density (VD)
The mass of the vapor divided by the volume it fills.

Solubility
A measure that indicates the tendency of a chemical to dissolve evenly in a liquid.

Once the surface tension is broken, the molecules are released in the form of vapor. The **open cup test** measures the release of the vapors in terms of the pressure exerted at a specific temperature. Temperature at the point where vapors given off will ignite is measured. The point at which a flash fire is produced when an ignition source is introduced into the vapor/air mixture above the liquid during heating is termed the **flash point.** The **fire point** closely follows. It is the temperature of a liquid that produces vapors that when heated will ignite the liquid and sustain burning. Further explanation of both flash point and fire point will be found later in this chapter.

In the open cup test, the flammable vapors or molecules driven off the flammable liquid are not trapped within an enclosure. The temperature at which the vapors are released is lower than the temperature recorded using a **closed cup test.**

In the closed cup test, place a lid on the cup and take the pressure reading when the vapors are released from the liquid. Open and closed cup ratings vary widely because of the pressure differences between the two test methods. In the closed cup test, a lid is placed over the sample, trapping the molecules driven off by the heat of the product. These molecules bounce against the surface of the lid, causing an increase in the pressure.

It is important to remember that the flash point varies depending upon the pressure, the oxygen content, and the purity of the product.

Vapor Density

Vapor density (VD) is the mass of the vapor divided by the volume it fills. Vapors with a vapor density greater than 1.0 will sink to the ground, and may accumulate and displace breathing air. Carbon dioxide is 1.5 times as heavy as air; therefore, a room filled with carbon dioxide will eventually have the lower portion of the room blanketed with vapors at the floor level. The oxygen in the room will be replaced or forced upward. Firefighters will become asphyxiated if the entire room fills with carbon dioxide.

Another concern is when the gas is flammable. For example, butane gas is twice as heavy as air and has a flammability range of 1.9% to 8.5%. This is of concern to firefighters as flammable vapors heavier than air may pool at the lowest point awaiting a source of ignition. The vapor density of butane coupled with its flammability range can present a very serious explosion threat. This is a good example of how an understanding of chemical properties is relevant to assessing risks on an emergency incident scene.

Solubility

Solubility in water indicates the amount of a material that will dissolve and mix in water. Insoluble or slightly soluble materials will form a separate layer and will either float or sink, depending on their specific gravity, and is thus termed

Insolubility

Insoluble or slightly soluble materials will form a separate layer and will either float or sink, depending on their specific gravity.

Polar solvent

Materials that are soluble in water.

insolubility. Solubility is sometimes described with words and sometimes with numbers; both indicate the percentage of the material that will dissolve.

The materials that are soluble in water are called **polar solvents.** Alcohols and other polar solvents dissolve in water so these materials may be diluted by firefighters to a point that they will not burn.

In many cases, nonpolar solvents or hydrocarbon liquids that are not soluble in water will float on top of water, presenting firefighters with a serious fire danger and safety problem. If they attempt to extinguish a fire in a hydrocarbon material floating on water, an ignition source can reignite the material, even after it has been extinguished, because the material floats on the surface of the water. It is a safety concern because it can encircle firefighters and reignite. The fire attack must be made from a location where firefighters cannot become surrounded with floating liquid that may reignite. Safety for firefighters can be improved by floating a thick, continuous coating of foam over the top of the material. (See Chapter 4 for details on the use and application of foam.)

Specific Gravity (Water and Air)

Specific gravity

The density of the product divided by the density of water, with water defined as 1.0 at a certain temperature.

Boil over

The expulsion of a tank's contents by the expansion of water vapor that has been trapped under the oil and heated by the burning oil and metal sides of the tank.

Liquefied petroleum gas (LPG)

A term given butane and propane gas that has been pressurized and contained in a tank.

Specific gravity is the density of the product divided by the density of water or air. Water and air density is defined as 1.0 at a certain temperature. Water and air are the standards that have been given the value of 1.0. Water is then measured against other liquids and air is compared against other gases. If the liquid being tested is heavier than water, it will sink to the bottom of the container. Firefighters fighting an oil tank fire must be aware of the possibility that the water applied by hose lines may sink to the bottom of the tank and heat up because the fire warms the metal sides of the tank and the oil. This heat from the metal tank walls and the oil itself is transmitted to the water trapped below the oil in the bottom of the tank. Once the trapped water is heated to 212°F (100°C), it is converted to steam and expands, blowing the oil upward and out of the top. This is called a **boil over** and is a very serious and dangerous safety problem for firefighters.

The specific gravity of air is defined as 1.0 at 70°F (21°C). Gases are then measured against the air to determine if the gas is either heavier or lighter than air at the same temperature. This means that the gases that are heavier than air seek the lowest elevation in a floor area. For example, of the two **liquefied petroleum gases (LPGs)** used, butane is 2.0 times heavier than air and propane is 1.5 times heavier, so firefighters can expect butane and propane vapors to seek the lowest areas and pool there. The fire code requires appliances with pilot lights or other sources of ignition to be located above the floor area as these heavier-than-air gases pool in low spots.

On the other hand, natural gas (methane) is lighter than air; it has a specific gravity 0.55 in relation to air, which is 1.00. This causes the vapors from methane to seek elevated locations in a confined room. Firefighters will need to ventilate the higher elevations of a confined space first, if possible.

Table 2-3 *Specific gravities of selected liquids.*

Substance	Specific Gravity at 68°F/20°C	Heavier/Lighter than Water
Acetic acid	1.05	Heavier than water
Allyl chloride	0.94	Lighter than water
Chlorobenzene	1.11	Heavier than water
Heptane	0.68	Lighter than water
Hydrochloric acid	1.19	Heavier than water
Nitric acid	1.50	1.5 times heavier than water
Sulfuric acid	1.84	1.75 times heavier than water

Table 2-4 *Common gases encountered by firefighters—heavier/lighter than air.*

Substance	Vapor Density (air = 1.00)	Heavier/Lighter than Air
Acetylene	0.899	Lighter than air
Ammonia	0.589	Lighter than air
Carbon dioxide	1.52	1.5 times heavier than air
Chlorine	2.46	Almost 2.5 times heavier than air
Hydrogen	0.07	Lightest of all gases
Methane	0.553	Lighter than air
Nitrogen	0.969	Lighter than air
Oxygen	1.11	Heavier than air
Propane	1.52	1.5 times heavier than air
Sulfur dioxide	2.22	2 times heavier than air

Knowing and understanding the specific gravity of the liquid or gas provides insight for firefighters to determine the best method of fire extinguishment or, in the case of a gas, the best and safest means of moving the gas to the open environment. Table 2-3 shows common materials and gases encountered by first responders. Table 2-4 shows examples of gases that may be encountered by first responders.

■ Note
The lower the flash point, the greater the fire hazard of a material.

Flash Point

As mentioned, the flash point is the minimum liquid temperature at which enough vapors are present above the liquid to ignite or flash but does not continue to burn.

Fire Point

The fire point is the lowest temperature at which a liquid produces a vapor that can sustain a continuous flame as opposed to the instantaneous flash of the flash point. Fire point is usually a few degrees above the rated flash point temperature. For example, the flash point of methanol is 52°F (11°C) while the fire point is 56°F (13.5°C).

The **auto-ignition temperature** is the temperature at which a material will ignite in the absence of any external source of heat. We refer to this temperature as the spontaneous ignition temperature, meaning that the material is heated to a point where it self-ignites.

Firefighters may encounter *tar pots* used to heat asphalt (to melt it for application on roofs). Once heated to its auto-ignition temperature, these pots will explode in fire when the lid of the pot is raised, allowing oxygen to mix with the superheated flammable gas. Water spray on the outside of the pot will cool it below the auto-ignition temperature, but getting water inside the pot will result in a violent reaction, splashing hot asphalt outside the pot.

Explosive Limits and Range

The **explosive range** is the range of concentrations of the materials in the air, which will permit the material to burn. The lowest ignitable concentration of a substance in air is called the **lower explosive limit (LEL).** The highest percentage of a substance in air that will ignite is the **upper explosive limit (UEL).** See Table 2-5 for a list of gases and their limits.

Acetylene has a wide explosive range (2.5% to 81%) and a BTU output of 1,499 per cubic foot, while hydrogen gas has a wider range (4% to 75%) with a lower BTU output of only 325 per cubic foot. Nevertheless, both present a serious fire and/or explosion hazard (NFPA 2000, Appendix A).

Table 2-5 *Flammability limits of materials encountered by firefighters.*

Material	LOWER % to Air	UPPER % to Air	Result
Acetone	2.6%	12.8%	Explosion or fire
Butane	1.9%	8.5%	Explosion or fire
Kerosene	0.7%	5%	Explosion or fire
Natural gas	6.5%	17%	Explosion or fire
Gasoline (92 octane)	1.5%	7.6%	Explosion or fire
Carbon monoxide	2.4%	74%	Explosion or fire

pH
A measure of a substance's ability to react as an acid (low pH) or as an alkali (high pH).

Acidity
A substance that will release hydrogen ions when dissolved in water.

Alkalinity
The amount of reaction of a substance with acids in aqueous solution to form salts, while releasing heat.

Hydrogen Ion Concentration (pH)

The **pH** of a chemical is a measure of **acidity** or **alkalinity,** where the number 7 is defined as the neutral point. The acidity or alkalinity of a substance is determined by the amount or concentration of hydrogen ions present (see Figure 2-2).

Appearance and Odor

Appearance and odor are properties that are important in the description of the material, including its color, smell, and physical state at normal temperature and pressure. If all properties are carefully considered, it is impossible to find two chemicals with identical properties.

Physical State

Hazardous materials can be liquids, solids, gases, or sludge that may be part solid and part liquid. Part of the risk assessment of a hazardous material at an incident

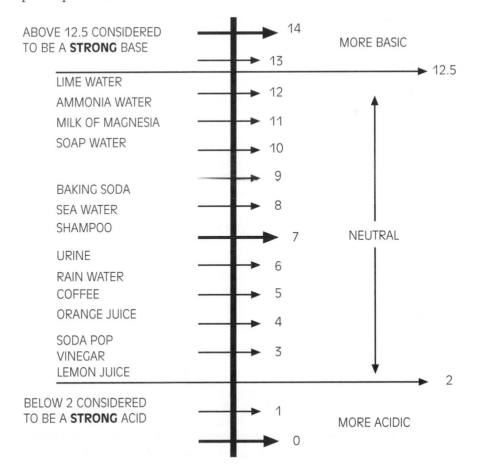

Figure 2-2 *pH values of common substances.*

■ Note

A product with a pH below 7 is acidic. If the pH is higher than 7, the material is basic or alkaline. The lower the pH, the stronger the acid; the higher the pH, the more corrosive the base. Materials with a pH of 2 or lower, or of 12.5 or higher, are classed as corrosive and are specially marked because of their danger to either people or the environment.

Basic

A substance that will react with acids in aqueous solution to form salts, while releasing heat.

Sublimation

The direct change from a solid to a gas without changing into a liquid.

Boyle's law

A theory that states the more a gas is compressed, the more the gas becomes difficult to compress further.

Charles's law

A gas will expand or contract in direct proportion to an increase or decrease in temperature.

depends on the physical state of the material. A spilled or leaked liquid will spread and seek out the nearest low-lying areas, while a spilled solid generally will form a pile, usually in the general vicinity of the spill. The configuration of the spilled material greatly influences the extent and hazard (threat) of the spill.

Solids The primary concerns with spilled solids on land are their flammability and air reactivity. If solids spill into water, the characteristics of solubility and reactivity must be considered. A few solids will form vapors without first becoming liquid, which is called **sublimation.** Examples of this are naphthalene (moth balls) and paradichlorobenzene (used as a solid bathroom deodorizer). The toxicity and flammability of the vapors must be assessed if subliming solids are spilled.

Liquids Liquids have the ability to flow away from a leak, thereby extending the hazard area. The viscosity of a liquid may be a factor in incident response, since the more viscous (thick and sludgy) liquids tend to flow more slowly and to stick to surfaces including the ground, clothing, and equipment. Sludge that has thickened due to evaporation may be concentrated and possibly more dangerous than the original liquid.

Gases There are some special considerations for emergency response to incidents that involve gases. Many gases are routinely shipped and stored in a variety of forms and containers, so it is likely that they will be encountered in incidents. Gases differ from liquids and solids in several ways, but the most unique characteristic is that they readily assume the shape and volume of their container. In most cases, liquids and solids are considered to be incompressible. Gases, on the other hand, are very elastic and the pressure exerted by the gas itself can measure this characteristic.

Boyle's Law The more a gas is compressed, the more the gas becomes difficult to compress further. This pressure created by compression within a container is described by **Boyle's law** of gases. It states that pressure of a gas is inversely proportional to its volume at a given temperature. This means that if the pressure is doubled in a specific volume in a closed container, the volume of the gas is reduced by one-half. For example, consider a tire that has 2 ft^3 of air. If that air is released into a 4 ft^3 container, it will have one-half the pressure than when it was in the 2 ft^3 tire.

Charles's Law Dr. Jacque Charles, a French scientist, found that a gas would expand or contract in direct proportion to an increase or decrease in temperature, so-named **Charles's law.** He found that the molecules of a gas are in continual motion; they collide with each other and the walls of their container. When these molecules are compressed, they bounce against the container walls with an even greater force. When the gas is heated, the movement of the molecules becomes faster which increases the pressure. If the walls of the container are elastic like a rubber balloon, the pressure will expand the outer walls of the balloon container. Likewise, if the gas is confined so that it cannot expand, its pressure will increase or decrease in direct proportion to the temperature.

Compressed Gases A **compressed gas** is any material that, when enclosed in a container, has an **absolute pressure** of more than 40 pounds per square inch (psi) at 70°F (21°C), or an absolute pressure exceeding 104 psi at 130°F (54°C), or both. Pressurized gases are liquefied when compressed and exist in a liquid–vapor relationship inside the containers.

Cryogenic liquids The **cryogenic** liquids are those materials with boiling points of no greater than –150°F (–101°C) that are transported, stored, and used as liquids. The super cold temperatures allow a large volume of gas to be stored as a liquid in a much smaller container at lower pressures. This super cooling decreases the needed storage space. The hazards of cryogenic gases relate to the nature of the particular gas, the large volume or ratio of vapor to liquid, and the extreme coldness.

One characteristic of cryogenic materials is their unique storage vessels. While the pressures are low (under a few psi), keeping a constant temperature is of utmost importance. Therefore, they are stored in cylinders designed like a thermos bottle. These containers have silver-coated linings (to reflect heat) and are designed to use the inherent temperature of the materials to help maintain the extremely cold environment. Because of the extreme cold, cryogens are capable of causing severe damage to anything that contacts the liquid.

Cryogenic liquids and liquefied gases vaporize rapidly upon release from their containers. A liquid spill will boil into a much larger vapor cloud. These clouds can be extremely dangerous, particularly if the vapors are flammable. Both cryogenic and liquefied gases result in freeze burns or severe frostbite requiring cold injury treatment procedures. Clothing saturated with cryogenic materials must be removed immediately. This is extremely important if the vapors are oxidizers or flammable, as the first responder cannot escape the flames from clothing steeped in the vapors if ignition occurs.

Cryogenic gases include argon, fluorine, helium, hydrogen, krypton, liquid natural gas, liquid neon, liquid nitrogen, liquid oxygen, and liquid xenon. Table 2-6 shows the boiling points of these substances and their expansion ratios.

Changes in Physical State

Some chemical reactions that occur during a leak or a fire cause a change in physical state that can lead to problems. A leaking liquid may change to the gas form with a tremendous increase in volume. In the vapor form, it may pool in lower areas or ascend depending upon its specific gravity. Containers heated in a fire may leak or explode due to an increase in the volume of contained gas or liquid. The expansion ratio of a chemical (the volume of its liquid form compared to the

Compressed gas
Any material that, when enclosed in a container, has an absolute pressure of more than 40 psi at 70°F, or an absolute pressure exceeding 104 psi at 130°F, or both.

Absolute pressure
The measurement of pressure exerted on a surface, including atmospheric pressure, measured in pounds per square inch absolute.

Cryogenic
The cryogenic gases are those materials with boiling points of no greater than –150°F (–65.5°C) that are transported, stored, and used as liquids.

Table 2-6   *Selected cryogens.*

Substance	Boiling Point (°F)	Expansion Ratio
Liquid argon	−302	840:1
Liquid fluorine	−306	980:1
Liquid helium	−452	700:1
Liquid hydrogen	−423	848:1
Liquid krypton	−243	695:1
Liquid natural gas	−289	635:1
Liquid neon	−411	1445:1
Liquid nitrogen	−320	694:1
Liquid oxygen	−297	857:1
Liquid xenon	−163	560:1

volume of its gas form) is known for all chemicals and is an important consideration in an incident involving the release of gases liquefied by pressure.

Combustible Dusts

Many materials in solid form may not combust; however, they can be made to combust or explode by converting them into dust. Under the right conditions, any organic dust can explode. In grain processing plants, dust explosions are not unusual. The finer ground the dust, the greater its potential for explosion. Confectioner's sugar and cornstarch are both finely ground, and as a result present the most severe explosion potential.

Many dust explosions occur in pairs. The first smaller explosion occurs near the vicinity of the ignition source. This first explosion lifts and scatters loose dust into the atmosphere, which ignites in a second, more devastating explosion, as more fuel is made available for ignition.

Many metals in a dust or shavings form will also explode. Magnesium, aluminum, and titanium are the more common metals firefighters encounter. A correctly designed and well-maintained fire sprinkler system with an excellent dust removal system and good housekeeping practices can reduce the risks of handling dust and decrease the potential of explosions.

Boiling Liquid Expanding Vapor Explosion

A **boiling liquid expanding vapor explosion (BLEVE)** occurs when a pressure tank has its container metal softened or weakened by heat or corrosion (see Figure 2-3).

BLEVE
The result of the explosive release of vessel pressure, portions of the metal tank, and a burning vapor cloud of gas with the accompanying radiant heat.

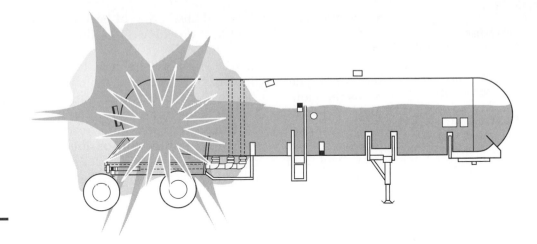

Figure 2-3 *BLEVE.*

When the expansion of vapors inside the container exceeds the pressure relief valve limit, the pressure causes the container to violently rupture into two or more pieces. All cylinders can be expected to fail (or exceed the pressure limit of the relief valve) at some point in time after being subjected to enough heating and expansion of the internal gas vapors.

BLEVEs most commonly occur when flames contact a tank shell above the liquid level or when insufficient water is applied to keep the tank shell cool. In the direct attack mode, the following points are important to protect the safety of the responders.

- In most cases these fires should not be extinguished, as it is generally safer for firefighters to control flames than to release gas vapors, which may result in an explosion.

- The goal of fire fighting operations is to cool the tank shell to a point that the liquid inside cools and reduces the amount of vapors and pressure being generated. If the pressure is kept below the pressure setting on the **pressure relief valve,** then the material will no longer be escaping the container and the hazard will be diminished.

- If control cannot be gained from hand lines cooling the vessel, then unattended monitor streams should be put in place for the safety of the firefighters.

CHEMICAL REACTIONS

The stability of a material is the result of strong **bonds** between atoms. The bonds are formed when atoms exchange or share electrons giving them the capacity to resist changes in a normal environment or exposure to shock or pressure, air,

Pressure relief valve
Used on compressed gas cylinders to release pressure buildup within a cylinder. Generally, the pressure at which they open is preset.

Bond (chemical)
The attractive force that often binds two atoms into a combination that is stable at least at room temperature, but which becomes unstable at a sufficiently high temperature.

Endothermic
The type of reaction in which energy is absorbed when the reaction takes place.

Exothermic
The type of reaction that will release or give off energy.

and water. This bonding and absorption of heat as well as the releasing of the bond with the release of heat are known as either **endothermic** or **exothermic** reactions.

An endothermic reaction is where heat (energy) is absorbed from the reaction or the "joining" of the molecules, which occurs when the bond that holds the molecules together is cemented. In Figure 2-4, the bond between two molecules of water (H_2O) is being formed with heat being absorbed.

When a bond is broken an exothermic reaction occurs, releasing heat (energy). Firefighters encounter this reaction on almost every fire situation where heat, light, and other products of combustion are released. In Figure 2-5, the bond between two molecules is being broken, thus releasing heat (energy).

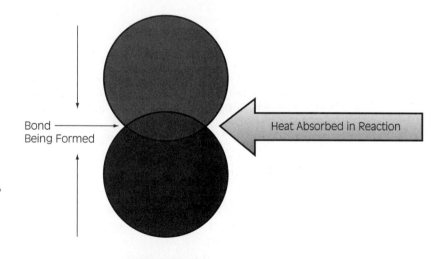

Figure 2-4 H_2O (the bond of water being formed).

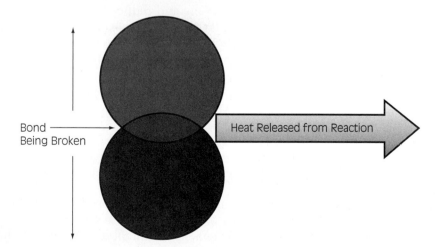

Figure 2-5
Exothermic reaction releasing heat.

It is important for firefighters to understand these chemical reactions as they often create very serious hazards for emergency responders and the environment. Some of the reactive materials which can cause problems for emergency responders are water-reactive materials, air-reactive materials, oxidizers, unstable materials, incompatible materials, and materials which **polymerize.** Polymerization is a process of reacting (linking or rearranging) monomer molecules together in a chemical reaction. The reaction can be explosive or extremely violent.

Polymerize
The process in which molecules of a monomer are made to combine with other monomers.

Water-reactive Materials

Water-reactive materials react with water, often violently, to release heat, a flammable or toxic gas, or a combination of the two. If there is a fire in which water is being used as the extinguisher, the presence of water-reactive materials can make the situation much more dangerous.

As an example, firefighters enter a metal disposal yard and attack a small fire in cardboard boxes, unaware that the boxes contain magnesium filings. They use the portable water-based extinguisher hanging on the wall in the storage area. Immediately they are faced with an explosion as the magnesium reacts with the water. One firefighter is seriously burned. This reaction is illustrated in Figure 2-6.

Air-reactive Materials

White phosphorous
This solid material is more dangerous than red phosphorus because of its ready oxidation and spontaneous ignition when exposed to air.

Air-reactive materials, as their name would imply, are reactive simply in the presence of air when they escape their containers. Some of the water-reactive materials are also air reactive and will ignite in air; potassium metal is one example. Diborane and trimethylaluminum are organic metal compounds which are air reactive as well. The air-reactive **white phosphorous** must be stored underwater to prevent its ignition.

Oxidizers

Oxidizers or oxidizing agents
Substances that present special hazards because they react chemically with a large number of combustible organic materials such as oils, greases, solvents, paper cloth, and wood.

Oxidizers, also called **oxidizing agents,** present special hazards because they react chemically with a large number of combustible organic materials such as oils, greases, solvents, paper, cloth, and wood. All organic and inorganic peroxides are highly combustible and some are highly reactive. The reaction is violent because the agents contain additional oxygen molecules and thus, when combined with the fuel, require only an ignition source for an explosive reaction to occur.

Some inorganic peroxides such as sodium peroxide and potassium peroxide are very reactive and sensitive to shock. Other commonly encountered oxidizers include ammonium nitrate, potassium permanganate, ammonium persulfate, and sodium nitrate. The violent reaction was illustrated in the 1995

Figure 2-6
Magnesium reacting to firefighters' hose line.

bombing of a federal building in Oklahoma City, Oklahoma. Figure 2-7 shows where ammonium nitrate, a powerful oxidizer, mixed with fuel oil was used as an explosive, resulting in the death of 286 people.

Another incident occurred in 1947, where the fire and resulting explosion of ammonium nitrate (an oxidizer) resulted in the death of 468 people, including all the first responding firefighters, in Texas City, Texas.

Unstable Materials

When exposed to water, air shock or pressure materials designated as unstable have a tendency to decompose, polymerize, or become self-reactive. One group of materials that is chemically unstable contains the **monomers,** or building blocks that form many types of polymers (resins, plastics, and synthetic rubber materials). Some of these materials can polymerize spontaneously, causing rupture of the container.

Monomers
Small molecules, usually gaseous or liquid, which are used to produce polymer resins.

Figure 2-7 *Federal building, Oklahoma City, Oklahoma.*

One example is spray foam insulation which is sold in an aerosol container at most hardware stores. It is used to seal walls and around windows. When the foam is released from the container, it reacts with air to expand. Once started, the reaction cannot be stopped; and if the reaction occurs in a container such as a railroad car, there is the possibility of a rapid and violent container failure.

Incompatible Materials

A review of basic chemistry for first responders would not be complete without mentioning the incompatibility of chemicals. Some materials, when mixed with other materials, can adversely affect human health and the environment in a variety of ways, taking the following actions.

- Generation of heat
- Violent reaction
- Formation of toxic fumes or gases
- Formation of flammable gas
- Fire or explosion
- Release of toxic substances if they burn or explode

Catalyst

Catalyst
A substance that greatly affects the rate of a chemical reaction, but is not created or destroyed in the chemical reaction.

Some reactions proceed more readily in the presence of a **catalyst.** A catalyst is a substance that is not created or destroyed in the chemical reaction, but greatly affects the rate of the chemical reaction itself. An everyday example of a catalyst is platinum, which is used in the catalytic converter of a car exhaust system. It causes the fuel to burn faster and cleaner, reducing the output of carbon residue without consuming the platinum.

TOXIC COMBUSTION PRODUCTS

A good example of chemical reactions and their hazard to firefighters is the toxic materials that form from reactions between chemicals during a fire and pose significant health hazards. Formation of these toxic products depends upon the nature of the burning material and the amount of oxygen present. For example, many home furnishings and decorations are made with plastics that when exposed to fire release toxic chemicals into the atmosphere. Wool and silk will give off hydrogen cyanide, a highly poisonous gas, under some circumstances. This is one of the many reasons why firefighters must wear respiratory protection or a **self-contained breathing apparatus (SCBA)** when conducting fire fighting activities, both during the fire and afterwards during the overhaul phase because gases are still present in the structure.

Self-contained breathing apparatus (SCBA)
Air is carried in a pressurized tank and provided to the person under a positive pressure within an enclosed mask.

It is important to note that chemical properties and chemical reactions are not limited to hazardous materials incidents. This important concept is relevant in everyday fire fighting activities. Fire itself is a chemical reaction. Combustion is a type of decomposition where a molecule breaks down to carbon and water with the evolution of heat and light. A solid understanding of chemical and physical properties of materials and chemical reactions is important for firefighters to have in nearly every task they perform. Chapter 10 provides a basic foundation in hazardous materials information systems and warning methods for first responders.

SUMMARY

Basic chemistry and physical processes are important to firefighters for emergency response actions and, more importantly, for their safety. Today, large amounts of hazardous chemicals are being transported by highway and are more often used in our daily lives. Homes and businesses contain a multitude of hazardous chemicals, and we have learned that combustion produces toxic by-products through chemical processes and reactions. To remain safe and get the job done, firefighters must be equipped with the knowledge and tools to investigate and determine the extent of the hazard. Then once they have identified the hazard they can actively take those actions required to control and mitigate the situation.

KEY TERMS

Absolute pressure The measurement of pressure exerted on a surface, including atmospheric pressure, measured in pounds per square inch absolute.

Acidity A substance that will release hydrogen ions when dissolved in water.

Alkalinity The amount of reaction of a substance with acids in aqueous solution to form salts, while releasing heat.

Atom Made up of a neutron that is electrically neutral, a proton that has a positive electric charge, and a surrounding cloud of orbiting electrons, which are negatively charged.

Auto-ignition temperature The temperature at which a material will ignite in the absence of any external source of heat. We refer to this temperature as the spontaneous ignition temperature, meaning that the material is heated to a point where it self-ignites.

Barometer An instrument for measuring atmospheric pressure.

Basic A substance that will react with acids in aqueous solution to form salts, while releasing heat.

BLEVE The result of the explosive release of vessel pressure, portions of the metal tank, and a burning vapor cloud of gas with the accompanying radiant heat.

Boil over The expulsion of a tank's contents by the expansion of water vapor that has been trapped under the oil and heated by the burning oil and metal sides of the tank.

Boiling point (BP) The temperature at which a liquid will convert to a gas at a vapor pressure equal to or greater than atmospheric pressure.

Bond (chemical) The attractive force that often binds two atoms into a combination that is stable at least at room temperature, but which becomes unstable at a sufficiently high temperature.

Boyle's law A theory that states the more a gas is compressed, the more the gas becomes difficult to compress further.

Catalyst A substance that greatly affects the rate of a chemical reaction, but is not created or destroyed in the chemical reaction. The catalyst helps to lower the energy activation thus increasing the reaction rate.

Cell(s) The objects that comprise all organisms, numbering one or more, where all vital functions of the organism occur within the walls.

Charles's law A gas will expand or contract in direct proportion to an increase or decrease in temperature.

Chemical change Molecules of a material are changed by a process.

Closed cup test Place a lid on the cup to confine the vapors above the cup. The point vapors are given off in sufficient quantities to ignite this point is called a *closed cup reading*. The ignition point is always lower in a closed cup test than an open cup test. The devices used to perform this test are the Tag Closed Cup Test for liquids below 200°F (93°C) and the Pensky-Martens Closed Cup Apparatus for liquids above 200°F (93°C).

Compound A substance formed from two or more elements joined with a fixed ratio.

Compressed gas Any material that, when enclosed in a container, has an absolute pressure of more than 40 psi at 70°F, or an absolute pressure exceeding 104 psi at 130°F, or both.

Cryogenic The cryogenic gases are those materials with boiling points of no greater than −150°F (−65.5°C) that are transported, stored, and used as liquids.

Electron A very light particle with a negative electrical charge, a number of which surround the nucleus of most atoms.

Element The simplest form of matter.

Endothermic The type of reaction in which energy is absorbed when the reaction takes place.

Exothermic The type of reaction that will release or give off energy.

Explosive range The range of concentrations of the gases or materials (dusts) in the air, which will permit the material to burn.

Fire point The lowest temperature at which a liquid produces a vapor that can sustain a continuous flame.

Flammable gas A gas that is flammable at atmospheric temperature and pressure in a mixture of 13% or less (by volume) with air that has a flammable range with wider than 12%, regardless of the lower limit.

Flammable range The numerical difference between a flammable substance's lower and upper explosive limits in air.

Flash point The temperature of a liquid which, if an ignition source is present, will ignite only the vapors being produced by the liquid creating a flash fire.

Functional groups Elements of the group that have the same electronic structure in the outermost shell, thus allowing the elements to show similar chemical behavior.

Halogenation or halogenated Any chemical reaction in which one or more halogen atoms are incorporated into a compound.

Inorganic A classification in which the matter consists primarily of the materials of the earth such as rocks, soil, air, water, and minerals in or below the surface of the earth.

Insolubility Insoluble or slightly soluble materials will form a separate layer and will either float or sink, depending on their specific gravity.

Liquefied petroleum gas (LPG) A term given butane and propane gas that has been pressurized and contained in a tank. Both gases are heavier than air and have higher heat content than natural gas. They are widely used in recreational vehicles and can be found in homes in the more rural areas where natural gas (or methane) is not available.

Lower explosive limit (LEL) The lowest ignitable concentration of a substance in air that will ignite.

Mass A measurement of quantity when the weight is proportional to the mass.

Matter Anything that occupies space and has mass or something that occupies space and can be perceived by one or more senses.

Molecule Two or more atoms tightly bound together by chemical bonds.

Monomers Small molecules, usually gaseous or liquid, which are used to produce polymer resins.

Negative ion Forces outside an atom cause it to gain an extra electron, giving it an overall negative charge. It now contains more negative electrons than protons, so the sum of the electrical charge is negative.

Neutron   A particle with nearly the same mass as the proton, but electrically neutral; it is part of the nucleus of all atoms except the most common isotope of hydrogen.

Open cup test   Flammable liquids are heated until the surface tension of the liquid is broken. At this breaking point, the fuel releases molecules in the form of vapor above the surface. The open cup test results in a higher ignition point than the closed cup test. The device used to perform this open cup test is called the Cleveland Open Cup Apparatus.

Organic   A classification in which the matter is found in living substances such as plants and animal life.

Oxidizers or oxidizing agents   Substances that present special hazards because they react chemically with a large number of combustible organic materials such as oils, greases, solvents, paper cloth, and wood. Halogens are also powerful oxidizers.

pH   A measure of a substance's ability to react as an acid (low pH) or as an alkali (high pH).

Physical change   Molecules of a material are not changed by a process.

Polar solvent   A substance that allows fire fighting foam to be used on alcohol-based fires without breaking down the soap-based materials in the foaming agent.

Polymerize   The process in which molecules of a monomer are made to combine with other monomers. Sometimes the reaction is explosive in nature.

Positive ion   An atom that is missing electrons.

Pressure relief valve   Used on compressed gas cylinders to release pressure buildup within a cylinder. Generally, the pressure at which they open is preset.

Proton   A positively charged particle that is the nucleus of the most common isotope of hydrogen.

Self-contained breathing apparatus (SCBA)   Air is carried in a pressurized tank and provided to the person under a positive pressure within an enclosed mask.

Solubility   A measure that indicates the tendency of a chemical to dissolve evenly in a liquid.

Specific gravity   The density of the product divided by the density of water, with water defined as 1.0 at a certain temperature.

Sublimation   The direct change from a solid to a gas without changing into a liquid.

Upper explosive limit (UEL)   The highest percentage of a substance in air that will ignite.

Vapor density (VD)   The mass of the vapor divided by the volume it fills.

Vapor pressure (VP)   The pressure placed on the inside of a closed container by the vapor in the space above the liquid which the container holds.

White phosphorous   This solid material is more dangerous than red phosphorus because of its ready oxidation and spontaneous ignition when exposed to air. It may be stored under water or oil.

REVIEW QUESTIONS

1. Define and give one example each of the three states of matter.

2. Describe the difference between a physical and a chemical change of a molecule.

3. Explain the difference between flash point and fire point.

4. Identify the major concerns for firefighters when faced with an incompatible chemical mixture.

5. Why is solubility critical when fighting a hydrocarbon fire?

REFERENCES

Delmar Cengage Learning. (2008). *Firefighter's Handbook: Firefighting and Emergency Response* (3d ed.) (Clifton Park, NY: Delmar Cengage Learning).

Department of Transportation. (2008). The Emergency Response Guidebook.

Faraday, Michael. (1993). *The Chemical History of a Candle* (Atlanta: Cherokee Publishing Company).

Friedman, Raymond. (1998). *Principles of Fire Protection Chemistry and Physics* (Quincy, MA: National Fire Protection Association).

Gantt, Paul. (2009). *Hazardous Materials: Regulations, Response & Site Operations* (Clifton Park, NY: Delmar Cengage Learning).

Hall, J. R. Jr. (1995). "Fire Risk Analysis: Model for Assessing Options for Flammable and Combustible Liquid Products in Storage and Retail Occupancies," *Fire Technology* 31(4), 290–308.

Hawley, Chris. (2008). *Hazardous Materials Incidents* (3d ed.) (Clifton Park, NY: Delmar Cengage Learning).

Hess, Fred C. (1996). *Chemistry Made Simple* (New York: Doubleday Publishing).

National Fire Protection Association. (2002). *National Fire Protection Handbook* (19th ed.) (Manchester, NH: NFPA).

Chapter

3

COMBUSTION PROCESSES

Learning Objectives

Upon completion of this chapter, you should be able to:

- Explain the theories underlying combustion processes.
- Describe how fire researchers have identified combustion processes using a variety of different classifications.
- Provide a description of the stages and events of fire as it progresses from the initial stage to its final stage.
- Explain the causes of flame over, flashover, and backdraft and review the procedures to prevent and protect against such events.
- Describe the various methods by which heat and unburned gases move in a confined environment.
- Define the five classes of fires and explain how they are classified.

INTRODUCTION

This chapter considers the physical and chemical process involved in fire combustion and relates these factors to the procedures the fire service has developed to confine, control, and extinguish uncontrolled fires. We will examine fires contained within rooms or structures with emphasis placed on these combustion processes as a fire progresses through the various stages and through the events of flame over, flashover and backdraft.

The chapter concludes with a review of the various fire classification methods, the fire extinguishing agents, and the advantages and disadvantages of these agents.

WHAT IS COMBUSTION?

Combustion is a planned and controlled, self-sustaining chemical reaction between a fuel and oxygen with the evolution of heat and light. The process is usually associated with the oxidation of a fuel by available atmospheric oxygen; however, some fuels may contain oxygen which is incorporated into the oxidation process. One definition of combustion is a process involving chemistry, thermodynamics, fluid mechanics, and heat transfer with a level of control in the oxidation of fuel.

This definition of combustion is not the same as the commonly used term **fire.** Fire differs from combustion because of the control of the combustion event itself. Another way to define combustion is to describe the process as the same chemical reaction as the slow rusting of iron or the same chemical reaction as an explosion of any flammable gas or substance. The only difference between the two is the speed of the reaction.

In the past, the fire service model of combustion was termed the **fire triangle,** representing three components needed for combustion to occur. The first component is an oxidizing agent, such as oxygen, which needs to be in the correct proportion (between 16% and 21% to air). The second component is a combustible material or fuel. The third component is an ignition source or heat energy source with sufficient heat or energy to raise the fuel to its ignition temperature. Figure 3-1 shows the first fire service model of combustion that was termed the fire triangle. It represents the three components needed for combustion.

The fire triangle model is a three-sided figure depicting the three processes of combustion (fuel, oxidizer, and heat). It has been replaced with a new graphic representation of combustion known as the **fire tetrahedron** (Figure 3-2). Research has found that the continual burning of a combustible material is governed in part by the heat (about 13%) of the gaseous flame that is transferred back into the fuel, causing the fuel to release **free radicals** which vaporize. This vaporization can occur with or without chemical decomposition of the

Combustion
A planned and controlled, self-sustaining chemical reaction between a fuel and oxygen with the evolution of heat and light.

Fire
A rapid, self-sustaining oxidation process that involves heat, light, and smoke in varying quantities.

Fire triangle
An older three-sided model used to describe the heat, fuel, and oxygen necessary for fire.

Fire tetrahedron
A four-sided model describing the heat, fuel, oxygen, and chemical reaction necessary for combustion.

Free radicals
A molecular fragment possessing at least one unpaired electron.

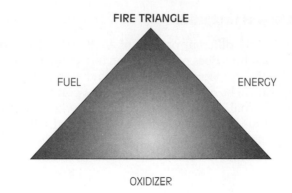

FIGURE 3-1 *The fire triangle.*

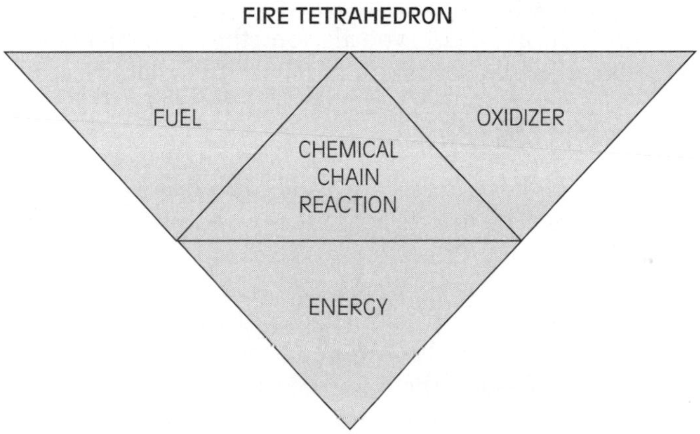

Figure 3-2 *The new fire tetrahedron.*

Pyrolysis
The process of breaking down a solid fuel into gaseous components when heated; also called thermal decomposition.

molecules. If chemical decomposition occurs, the process is called **pyrolysis.** Once started, the burning will continue until one of the following events occurs.

- The combustible material is consumed.
- The oxidizing agent concentration is lowered to below a level necessary to support combustion.
- Sufficient heat is removed or prevented from reaching the combustible material, thus preventing further fuel pyrolysis.
- The flames are chemically inhibited or sufficiently cooled to prevent further reaction.

This finding led to the development of a new model in order to depict the fourth process. A fire tetrahedron is a four-sided figure depicting the four processes of combustion and incorporates the addition of the uninhibited (self-sustaining) chemical reaction. Any one side of the figure being removed will result in the fire being extinguished.

Spontaneous combustion
An occurrence where a material self-heats to its piloted ignition temperature, then ignites.

Piloted ignition temperature
The ignition temperature of a liquid fuel. When heated, it will self-ignite.

Particulates
The unburned products of combustion visible in smoke.

Entrainment
The gathered or captured cooler air that replaces the rising heated air surrounding the point of combustion.

Smoldering (glowing) combustion
The absence of flames with the presence of hot materials in the surface where oxygen diffuses into the surface of the fuel.

Spontaneous Combustion

One slightly different process of combustion is **spontaneous combustion,** or auto-ignition, which does not require an independent ignition source. In the spontaneous combustion process, the combustible material itself heats to its **piloted ignition temperature** before it is ignited. At this point, it begins to support flame spread. For example, coal is a porous solid material where air can enter and diffuse into the interior of the material, but the heat produced within the stored coal is contained by the insulation properties of the coal. As the coal begins heating because of the trapped heat, it eventually reaches its ignition temperature and combustion begins.

METHODS OF FIRE CLASSIFICATION

Fire has been classified in five ways: the type of combustion, the rate of fire growth, available ventilation, the type of materials that are burning, and the stages or phases of a fire. These five classifications are detailed in this section.

Types of Combustion

One of the simplest methods of classification is the stages of the combustion process. The three stages are precombustion, smoldering combustion, and flaming combustion. These three stages do not necessarily occur in sequence (Delmar Cengage Learning, 2008).

When fuels are heated to their ignition point, the process is best defined as precombustion. The applied heat causes the fuel to release vapors and **particulates.** Particulates are the unburned products of combustion that are seen as smoke. As the efficiency of the combustion process is improved with additional heat and oxygen, the diameter of the particulates becomes smaller. In some situations, because of their small size, a point is reached when the smoke (particulates) is almost invisible to the human eye.

The precombustion process continues in two ways. First, the heat raises the fuel to its ignition temperature, releasing vapors which then combust. Second, the combustion process releases hot gases, which rise above the fuel and in doing so, **entrainment** gathers additional oxygen from the surrounding environment. A small portion (approximately 13%) of the heat energy is radiated back into the fuel, which in turn releases a stream of unburned vapors to keep the process flowing in a continuous cycle. This cycle, if not broken by firefighters, will continue until the fuel is consumed. This process is illustrated in Figure 3-3.

Smoldering Combustion

Smoldering (glowing) combustion is distinguished by the absence of flame and the presence of hot materials on the surface where oxygen diffuses into the fuel. It may or may not be related in any way to the oxygen content in the area. The

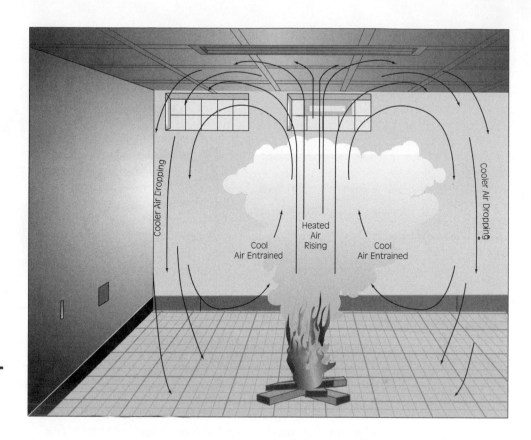

Figure 3-3 *Fire plume with cooler air being entrained.*

Within the figure: Cooler Air Dropping · Cooler Air Dropping · Cool Air Entrained · Heated Air Rising · Cool Air Entrained

glowing is indicative of temperatures in excess of 1,000°F (537.8°C). Smoldering combustion involves two phases: solid and gas. In the first phase, the solid material (fuel) is converted into the second phase, the gas phase, which, because of the conversion of the solid to gas, allows the combustion process to continue at or near the surface. In Figure 3-4, three examples of smoldering (or glowing) combustion are distinguished by the absence of flame and the presence of hot materials on the surface where oxygen diffuses into the surface of the fuel.

Smoldering combustion is a deadly process as the incompleteness of the combustion process creates very high levels of **carbon monoxide (CO)**. During this process, more than 10% of the fuel mass is converted into CO. This colorless, tasteless, and odorless gas is present at every fire. The problem presented by CO to firefighters is that the body has an affinity for it. In the lungs, CO competes with oxygen to bind with the hemoglobin molecule. Hemoglobin prefers CO to oxygen and accepts it over 200 times more readily than it accepts oxygen. In addition to the high body affinity, it is flammable and has a very wide flammable range of 12.5% to 74%. Its ignition temperature is 1,128°F (609°C). Other toxic gases can be found in smoke as well. They are displayed in Table 3-1.

Carbon monoxide (CO)
The result of the incomplete combustion of a fuel where carbon and a single atom of oxygen are produced.

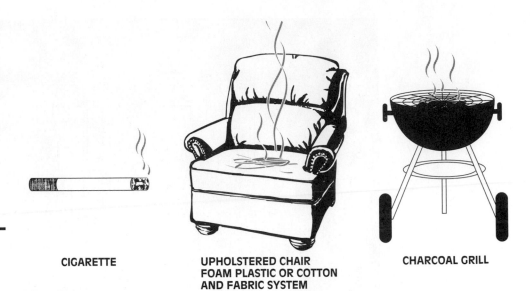

CIGARETTE

UPHOLSTERED CHAIR
FOAM PLASTIC OR COTTON
AND FABRIC SYSTEM

CHARCOAL GRILL

Figure 3-4 *Three examples of smoldering combustion.*

Table 3-1 *Toxic fire gases.*

Acrolein (C_3H_4O) is a strong respiratory irritant that is produced when polyethylene is heated and when materials containing cellulose, such as wood and other natural materials, smolder. It is used in the manufacture of pharmaceuticals, herbicides, and tear gas.

Hydrogen chloride (HCl) is a colorless but pungent and irritating gas given off during the thermal decomposition of materials that contain chlorine such as polyvinyl chloride (PVC) and other plastics.

Hydrogen cyanide (HCN) is a colorless gas with a characteristic almond odor. Twenty times more toxic than CO, it is an asphyxiant and can be absorbed through the skin. HCN is produced in the combustion of natural materials, such as wool and silk containing nitrogen. It is also produced when polyurethane foam and other materials that contain urea burn. The concentrated bulk chemical is also used in electroplating businesses.

Carbon dioxide (CO_2) is a colorless, odorless, and nonflammable gas as produced in free burning fires. Although it is nontoxic, CO_2 can asphyxiate by excluding the oxygen from a confined space. It can also increase a person's intake of toxic gases by accelerating his or her breathing rate.

Nitrogen oxides (NO_2 and NO) are two toxic and dangerous substances liberated in the combustion of pyroxylin plastics. Because nitric oxide (NO) readily converts to nitrogen dioxide (NO_2) in the presence of oxygen and moisture, NO_2 is the substance of most concern to firefighters. Nitrogen dioxide is a pulmonary irritant that can also have a delayed systemic effect. The vapors and smoke from the oxides of nitrogen have reddish brown or copper color.

Phosgene ($COCl_2$) is a highly toxic, colorless gas with a disagreeable odor of musty hay. It may be produced when refrigerants such as Freon contact flame. It can be expected on fires in cold storage facilities. It may also be a factor in fires in heavy-duty heating, ventilating, and air-conditioning systems. It is a strong lung irritant, the full deleterious effect of which is not evident for several hours after exposure.

Figure 3-5 *Flaming combustion is the phase firefighters will most likely encounter.*

The ratios of oxygen and fuel are not an issue for the combustion process, as the chemistry of the reaction still determines the total consumption of each. The proportions of fuel, oxygen, and end products are called the **stoichiometry of reaction. Stoichiometric,** or ideal, reactions are perfectly balanced reactions which result in the almost complete consumption of all materials with little or no energy waste.

Flaming Combustion

Flaming combustion is the phase that most firefighters will encounter in an emergency incident. Flaming combustion can be identified by the presence of flames (see Figure 3-5). There cannot be flaming combustion *unless a gas or vapor is burning.* For a better understanding of the process, flaming combustion can be broken into two categories.

The first category is when the gaseous fuel is premixed with air before ignition. As an example, this combustion process can be found in a number of appliances (water heaters, gas furnaces, etc.) where hydrocarbon fuel is premixed with air in a specially designed chamber to produce a flame for heating purposes. It is accomplished with a mixing chamber that mixes the fuel and air that is taken into the chamber.

On most appliances, the chamber can be adjusted so the mixture is proportioned to burn correctly. The device is designed in an attempt to create a stoichiometric reaction where almost complete consumption of the fuel occurs. Generally the fuel, when burning, will be blue. Incorrect adjustment of the

Stoichiometry of reaction
The proportion of fuel to oxygen and the resulting end products.

Stoichiometric
An ideal burning situation or a condition where there is perfect balance between the fuel, oxygen, and end products.

Flaming combustion
An exothermic or heat-producing chemical reaction with flames occurring between a substance and oxygen.

Diffusive flaming
A combustion process where the flames are part of the actual process.

Class A fires
Fires involving ordinary combustibles.

Cellulosic materials
Materials produced by the chemical modification of cellulose.

Class B fires
Fires involving flammable liquids.

Class C fires
Fires involving energized electrical equipment or wires.

device may result in the production of the deadly gas CO, which is the product of an incomplete combustion process where the unburned carbon causes the flames to appear yellow.

The second category is **diffusive flaming.** The diffusive flaming process is characterized by flames, generally yellow in nature as the burning process is not complete. It is as if it were occurring in an incorrectly adjusted mixing chamber. It is the most common type of flaming that will be encountered by firefighters. In this case, the flame is actually part of the combustion process itself, as it proceeds with the production of gaseous materials. In addition to the gaseous materials, light and heat are emitted as well.

Firefighters need to be aware that the yellow flame indicates that CO is being produced. Most of these materials, when burning, will produce a yellow to orange flame, and an unusual color will be produced when a material such as alcohol burns, as it burns with a blue flame. Although flame color is generally related to the oxygen mixture and the fuel itself, temperature does impact flame color under certain conditions. Under laboratory conditions, flame colors have been observed from dark red to white, with colors ranging from a low of 930°F (500°C) as a dark red color to a white color with a temperature of 2,550°F (1399°C).

Fire Classification by Type of Substance Burning

Another method to classify fires is to identify them by the type of substance that burns and, in doing so, identify the most effective extinguishing agent for that particular classification. Figure 3-6 shows a list of the fire classifications. These classes are A, B, C, D, and K.

- **Class A fires** result from the combustion of ordinary **cellulosic materials** such as wood, paper, and similar materials including petroleum products such as rubber and plastics. Class A fires should burn with an ember or leave ash particles. Class A fires are typically extinguished by using the cooling effect of water.

- **Class B fires** are in flammable and combustible liquids or flammable gases. In most cases, they are extinguished by removing the oxygen supply and cooling.

- **Class C fires** result from the combustion of insulation materials in energized electrical equipment. Water is not recommended as it will conduct electricity and may result in an electrical shock unless the electricity can be safely deenergized.

- **Class D fires** are in combustible metals such as titanium, magnesium, aluminum, and sodium. Class D fires often require the use of special agents designed to extinguish fire in a specific metal. Firefighters inspecting these locations should confirm that the proper type of extinguishing agent is located near the metal being processed. While water

Class D fires
Fires involving combustible metals.

Class K fires
Fires involving cooking oils.

Saponification
The process of chemically converting the fatty acid contained in the cooking medium to soap or foam.

can be used on some Class D fires, its application must be applied only after a careful verification of the safety of its use.

- **Class K fires** is a recently added classification. Class K fires involve cooking oils and fats. These fires burn very hot, and in certain situations, dry chemical extinguishing agents are not the most effective extinguishing agent. For example, in applications such as kitchen ranges, hoods, ducts, and deep fat fryer fire protection, the extinguishing mechanism for both dry and wet chemicals is based on the process of saponification.

Saponification is the process of chemically converting the fatty acid contained in a cooking medium (oil or grease) to soap or foam. The soap or foam produced by this process is readily broken down by exposure to heat. By adding air and water, the soap or foam forms bubbles of air contained within a water-soap membrane. The bubbles can be made lighter (more air) or wetter (more water) depending on the type of coverage needed by the fuel burning. The foam then can be applied over the top of a burning fuel to cool and smother. The extinguishment process using foam made from the saponification process can be more effective in special application locations where water or dry chemicals are not effective.

Fire Classification by Stages and Events

The process of combustion in enclosed compartments or rooms occurs in four sequential stages. Some fires may display one or more of the three events described below in addition to the four stages. Firefighters will be better prepared to understand the burning process if they are able to recognize the various stages and events which may occur during fire development and will be better able to efficiently and safely extinguish the fire.

The fire stages are as follows:

- Ignition stage
- Growth stage
- Fully developed stage
- Decay stage

Ordinary Combustibles   **Flammable Liquids**   **Electrical Equipment**   **Combustible Metals**   **Combustible Cooking**

Figure 3-6 *Five classifications of fire.*

Incipient or ignition stage
The point at which the four components of the fire tetrahedron come together and the materials reach their ignition temperature and the fire begins.

Growth stage
The stage where the fire increases its fuel consumption and heat generation.

Fully developed stage
The stage of a fire where the maximum generation of heat and consumption of fuel and oxygen occur.

Flame over or rollover
The flames that travel through or across unburned gases in the upper portions of the confined area during the fire's development.

Flashover
A sudden event that occurs when all the contents of a room or enclosed compartment reach their ignition temperature almost simultaneously, resulting in an explosive fire.

The fire events are as follows:

- Flameover or rollover
- Flash over
- Backdraft

The first stage (the **incipient or ignition stage**) occurs after ignition. The flames are small and are contained within the materials of first ignition. The oxygen supply is not reduced and is approximately 21% (in the atmospheric air). There is a plume of hot gases rising from the flames with flame temperatures as high as 1,800° to 2,200°F (982.2° to 1,204.4°C). The heated convection currents carry the products of combustion to the upper parts of the room, drawing in additional oxygen at the bottom of the flames to sustain the combustion process. This spreading of the fire begins to move the fire to the second stage.

As the fire grows in intensity, more fuel becomes involved, releasing additional heat energy. The fire enters into the second stage at this point. The second stage is described as the **growth stage** where the fire is accompanied by increased fuel consumption and heat generation. The convection currents carry the heated gases to the upper part of the room where, upon reaching the ceiling area, they spread horizontally until they reach the walls.

After reaching the walls, the heated gases begin the bank downward into the lower portion of the room (Figure 3-7). These same gases radiate heat to the remaining parts and contents of the room. This spreads and increases heat intensity, which brings the materials in the room up to their ignition temperature. At this point, the fire is at the **fully developed stage.** During this stage the oxygen in the room is eventually consumed and reduced to a percentage between 9% and 12%. The temperatures in the upper portions of the ceiling reach 1,000°F.

The reduction in oxygen begins to reduce the speed of the combustion process; however, the upper room levels continue to become loaded with superheated fuel–rich products. The floor area is still relatively cool and still contains oxygen; a hot layer zone and a cool layer zone exist in the fire environment. The upper layer of gases is hot and under some conditions contains a large concentration of unburned particles. If the heated layer of gases reaches ignition temperature of the unburned heated particles, then a **flame over** or **rollover** event will occur. (See the accompanying box for a detailed explanation of fire events.) If a flame over does not occur, and the heat intensity increases, then the room contents in the lower cool zone begin to heat up toward their ignition temperature. Once they reach their ignition temperature, a **flashover** event occurs.

If a flashover event takes place, it will occur during the transition between the *growth phase* and the *fully developed stage.* In this stage, the fire is

Backdraft
A sudden, violent reignition of the contents of a closed-container fire that has consumed the oxygen within the space when a new source of oxygen is introduced.

generating its maximum heat and flames while consuming the oxygen and fuel in the compartment.

As the fire consumes the available fuel and oxygen in the compartment, the atmosphere remains hot but the rate at which the heat is being released begins to slow as the oxygen in the compartment is consumed. Large quantities of unburned gases and particles are given off. At this point, because the oxygen has been consumed and the compartment container is still very hot, the introduction of oxygen into the environment results in the **backdraft,** or smoke explosion, event.

However, if firefighters vent the fire under controlled conditions, the venting allows the release of the confined fire gases. This will prevent a backdraft; however, the fire will continue but at a much slower rate. The fire is then at the **decay stage.** The remaining mass of glowing embers continues to produce heat for a period of time, but eventually all the fuel will be consumed and combustion will cease.

Decay stage
The stage when the fire has consumed all of the available fuel and the temperature begins to decrease as the fire reduces in intensity.

Figure 3-7
Temperatures associated with the stages of fire and the unique fire events.

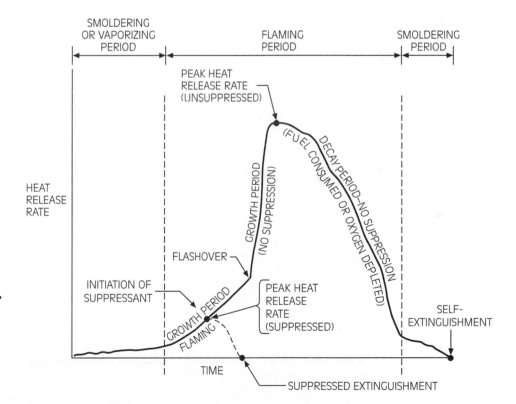

Fire Events

Fires in compartments or containers, such as rooms and buildings, can become deadly as the conditions change during the combustion process. Let's examine these fire events in more detail.

Flame Over

Flame over, also known as rollover, is a situation where flames travel through or across the unburned gases in the upper portions of the confined area during a fire's development, as shown in Figure 3-8. Rollover differs from a flashover because only the fire gases are involved, not the surfaces of other fuels within the confined space. Generally, the situation occurs during the growth stage as the hot gas layer accumulates at the ceiling or upper portion of the compartment.

If the mixture of the fuel and oxygen are within the explosive range, an explosion will follow, which will increase the interior pressure of the confined space. This increased pressure can result in the breakage of windows, or the destruction of the walls, doors, and window areas of the room. Firefighters trapped in this inferno may be overcome by the intense rapid increase in heat. The best action is to quickly get to the floor area in an effort to let the fire blow over, and then crawl to the exterior of the confined space.

Flames may be observed in the heated gas layer as the combustible gases reach their ignition temperature. The flame over phenomenon has also been called the "dancing of angels." Although the flames contribute to the total

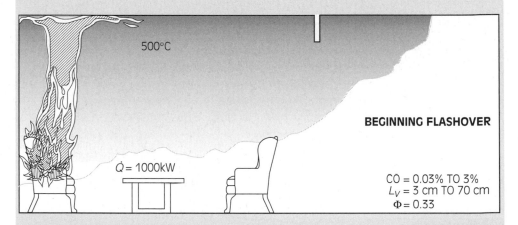

500°C

$\dot{Q} = 1000 \text{kW}$

BEGINNING FLASHOVER

CO = 0.03% TO 3%
L_V = 3 cm TO 70 cm
Φ = 0.33

Figure 3-8 *Flame over/rollover.*

(continued)

heat generated in the confined space, this condition is *not* the situation defined as flashover, but it can be an indicator that flashover is imminent.

Flame over may also be observed when unburned gases vent from a confined space during the fire's growth. These hot gases vent from the burning area of the confined space into the adjacent space. In doing so, they mix with oxygen and if they are at their ignition temperature, flames often will become visible in the gas layer.

Flashover

The condition prior to a flashover is where open flames gradually become less prevalent. In this period, the oxygen level drops below 16%, and while open flame still exists, the combustion process decreases, even in the presence of unburned fuel. In some cases it may even cease completely. At this point, if the fire is ventilated and the combustion process slows, the predominant combustion process becomes smoldering combustion, which may continue as long as there is sustainable fuel. Highly heated gases begin to build up at ceiling level. If this heating is enough to bring other materials in the room to their ignition temperature, all fuel materials in the room will ignite into flaming combustion. This phenomenon is called a flashover, as shown in Figure 3-9, which is a very dangerous event for firefighters.

■ **Note**
Firefighters need to know that they cannot survive a flashover because the entire confined space is spontaneously ignited; all materials have been brought up to their ignition temperature.

■ **Note**
In a situation where flashover is imminent, firefighters have two options: If water is available, they should immediately open the nozzle and cool the interior to below the ignition temperature of the materials. If this is not possible, they must get out of the closed environment immediately.

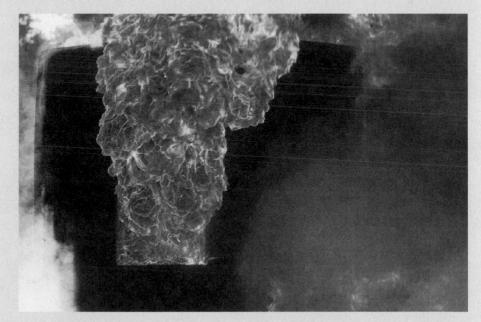

Figure 3-9   *Flashover.*

(continued)

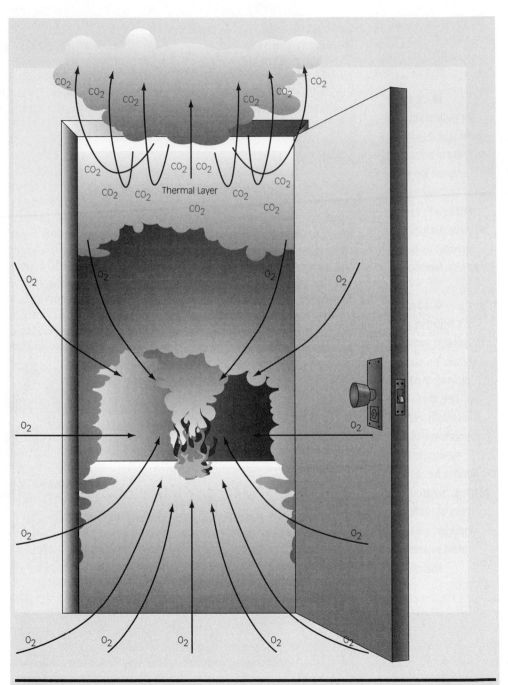

Figure 3-10 *Backdraft conditions.*

(continued)

■ **Note**

When a backdraft occurs, the additional oxygen entering the compartment is heated and expands, increasing the pressure inside the room which causes windows, walls, and weak points in the building to be suddenly pushed outward. Firefighters caught in this sudden, explosive rush of fire can be killed instantly.

Backdraft

Combustion is a balancing act between the air, the fuel, and the rate at which the fuel is being consumed. In this process, if the ventilation is limited, the fire will progress at a slower rate and the temperature will rise gradually. At the same time, the production of smoke or unburned products of combustion will be increased leading to a room or confined space filled with the products of incomplete combustion.

If the fire burns until the oxygen in the room is significantly reduced, the fire will appear to breathe as it draws in additional oxygen around cracks and small openings in the buildings. As the drawn-in oxygen increases the combustion process, it only slightly raises the pressure in the confined room which pushes out the smoke, thus relieving the pressure and the cycle starts over again. This "breathing" fire is awaiting a large influx of oxygen that will be provided by firefighters opening a window or door. Once this opening is made, the rush of oxygen into the room leads to a smoke explosion or a backdraft (see Figure 3-10).

Backdraft can be prevented by proper ventilation techniques. A vented opening allows the hot pressurized gases to be released to the exterior of the building, which reduces room temperature and smoke and allows firefighters to direct the movement of fire. Most importantly, the pressure of the confined gases is relieved. See Chapter 6 for a discussion on special concerns in ventilation procedures used to prevent backdrafts.

BUILDING CONSTRUCTION AND FIRE SPREAD

The efficiency of fire fighting forces declines quickly if the fire moves vertically through the building or bypasses horizontal construction barriers. For example, pre-World War II high-rise construction included thick, concrete walls, floors, and ceiling areas with columns on 30-ft centers with windows and few or no vent shafts for air conditioning and other building service amenities. The concrete walls and columns would absorb a great deal of the heat while the windows could be used to release heat and smoke to the outside environment. Thus, horizontal movement of the fire was inhibited by the presence of thick, solid, heat-absorbing concrete.

High-rise buildings built since World War II are built with truss construction and with drywall in the interior to provide fire resistance for the steel support structure of the building. Drywall does not have as much ability to absorb heat as thick concrete and, as a result, much of the heat from the fire is released back into the environment of the room. This quick buildup of heat within the room spreads the heated fire gases into other portions of the building. The fire gases quickly move to other portions of the building where they threaten occupants who are attempting to exit the building.

Compartmentation
Generally used in high-rise buildings to provide a safe area within the structure where persons can safely be protected from fire.

Compartmentation, or construction that includes safe areas, is a concept that proposes surrounding specific areas of buildings. For example, one floor out of every twenty floors of a high-rise should be equipped with four-hour rated, fire-resistive materials. An example of compartmentation would be having these areas built to serve perhaps twenty floors of occupants so that during a fire, the occupants from the twenty floors above the fire floor would evacuate to the safe area where the fire-resistive construction and opening protection devices would hold off the fire until fire fighting forces could control the fire movement. This is one of several proposed solutions to the problem of protecting persons trapped in these lightweight constructed buildings.

Fire Rating of Materials

There are also differences in a building's ability to withstand a fire because of variations in the workmanship on the materials or the way they are installed, different sets of test methods, or sizes of the test specimens. Small test specimens may be too small and not representative of the actual material when used in the building. Most firefighters will agree that the rated fire resistance of construction has some but not a substantial impact on the spread of fire.

Weather Conditions

Hot, dry, windy conditions impact the burning characteristics of inside building fires as well as outside fires in a number of ways. In fires inside rooms or confined areas, the hot, dry air works to dry out furniture, wall decorations, and ceiling coverings. This added heat from the weather raises the potential fuel closer to its ignition temperature by driving the moisture out of the fuel.

Stack effect
The temperature difference between the outside temperature of the building and the temperature inside the building.

The outside weather also impacts the movement of air currents in high-rise buildings with what is known as the **stack effect,** the temperature difference between the outside temperature of the building and the temperature inside the building. In other words, the interior of a high-rise building has air currents similar to ocean currents moving in patterns driven by temperature differences, which can spread the movement of heated fire gases throughout these buildings.

Windy conditions outside the building can impact horizontal ventilation activities if windows are used to remove fire gases. Firefighters need to make sure that the opening of windows does not spread the fire into other building areas not otherwise impacted by the fire.

Relative Humidity

Relative humidity
The ratio of the amount of moisture in a given volume of space to the amount that volume would contain if it were saturated.

Moisture in the form of water vapor is measured as **relative humidity.** It is always present in the air. The amount of moisture in the air affects the amount of moisture in the fuel; this applies to both indoor and outdoor fuels. The dry air with low moisture content will absorb water from the fuels. Likewise, near the ocean or other water source, dry fuel will absorb moisture from the air. The moisture content of the air and fuel is important in fire fighting, as drier fuels will ignite and burn much more rapidly and burn cleaner and hotter and with less smoke or less unburned particles than wetter fuels. Warehouse fires containing furniture that has been stored for long periods of time is a good example of where the drying effect of storage can impact the burning characteristics of the fuels. After extended storage time, the furniture tends to become dried out, which lowers its moisture content, and in turn lowers its ignition temperature.

This same burning characteristic can be seen in outside vegetation fires where the air layer near the ground influences both the fire and fuel behavior. If the relative humidity is 30% or below, the fire will burn freely. If the relative humidity is at 10% or below, the fire danger is critical, and extreme fire behavior (or behavior where the fire spreads very rapidly and in unpredictable directions) will occur, as the fuels are dried and preheated.

On the other hand, high humidity or a high percentage of moisture in the air also impacts the direction of fire gas movement. Higher humidity means the air is heavier (more water moisture) and heated fire gases tend to move horizontally, spreading the fire outward.

Mass/Drying Time

Mass, or the quantity of material, has an impact on how long it will take a source of ignition to raise the material to its ignition temperature. The thicker or heavier the mass, the longer it will take to raise the temperature of the material. This principle, when applied to the heating of water droplets, explains why smaller water droplets complete the conversion from water to steam more rapidly than larger water droplets. The smaller and less dense the water, the faster it is raised to its conversion temperature. In scientific terms this is identified as the **law of latent heat of vaporization.** It tells firefighters how much heat will be absorbed by the substance when it changes from a liquid (water) to a vapor (steam). Firefighters use the principle of the latent heat of vaporization concept to convert water to steam, allowing the steam to absorb the heat of the fire gases from the building and fire areas and move it to the exterior of the building.

Law of latent heat of vaporization
The heat that is absorbed when one gram of liquid is transformed into vapor at the boiling point under one atmosphere of pressure and the result is expressed in BTUs per pound or calories per gram.

HEAT MEASUREMENT

The temperature of a material is the condition that determines whether heat will transfer to or from other materials. *Heat always flows from higher temperature materials to lower temperature materials.* Temperature is measured in degrees using four scales, two of which are mainly used in research laboratory settings. The four scales used are:

- Kelvin (mainly research);
- Rankin (mainly research);

- Celsius; and
- Fahrenheit.

A comparison of these scales is found in Figure 3-11.

HEAT TRANSFER

The transfer of heat is important in all aspects of the combustion process from ignition through final extinguishment as its presence is responsible for the continuance of the combustion process as it releases the vapors and free radicals from the fuel. It is important for firefighters to understand how heat is transferred because fire extinguishment is the interruption of the transfer process. Heat is transferred by one or more of four mechanisms.

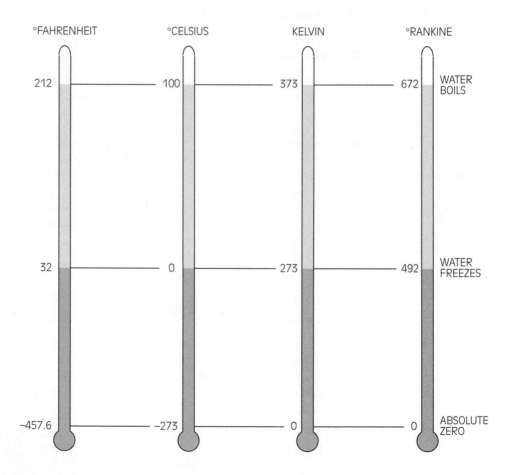

Figure 3-11
Relationship among temperature scales.

1. Conduction
2. Convection
3. Radiation
4. Direct flame impingement

Conduction

Conduction The transfer of heat energy from the hot to the cold side of a medium by means of energy transfer from molecule to adjacent molecule, or atom to atom.

Conduction is the transfer of heat energy from the hot to the cold side of a medium by means of energy transfer from molecule to adjacent molecule or atom to atom (see Figure 3-12). Its impacts are most noticeable in solid materials where molecular contact is at its highest and where convection air currents do not occur. Heat always travels from the hotter areas of a solid to the colder areas.

One way to think of heat transfer is to visualize the heated material containing moving molecules which, when in contact with a colder material (where the molecules are not moving as fast), transfers this movement by the touching of the molecules. Thus, a transfer of energy occurs. For example, heat traveling by conduction through a rod will eventually raise the temperature at the other end of the rod. The amount of heat flowing through the rod is proportional to the time, cross-sectional area, and difference in temperature between the ends; it is inversely proportional to the length.

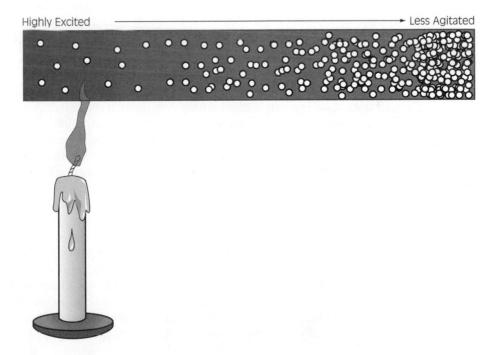

Figure 3-12
Conduction is the transfer of heat energy from a material by direct contact between the movements of molecules of another higher energy material.

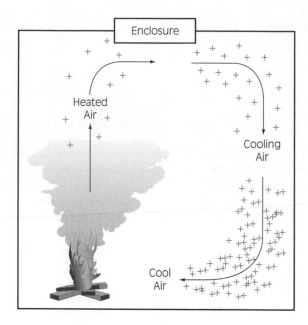

Figure 3-13
*Convection involves
the transfer of heat
by circulating
currents.*

Convection
The movement of heat
energy by the agitation
of air molecules that
reduces the density of
molecules, making
heated air lighter
than cooler air.

Radiation
The combined process
of emission,
transmission, and
absorption of heat
energy traveling
through
electromagnetic
waves from an area of
higher heat energy to
an area of lower heat
energy.

It is important for firefighters to be aware of thermal conductivity, or the rate at which heat is transmitted through a material, as it provides an estimate of time that the material will be able to resist the effects of a fire.

Convection

Convection involves the transfer of heat by a medium such as circulating air currents. For example, heat generated in a stove is distributed throughout a room by heating the air in contact with the stove (by conduction across the stationary boundary layer in contact with the surface of the stove). The hot, buoyant air rises, creating convection currents that transfer heat to distant objects in the room. Heat is transferred from the air to these distant objects again by conduction across the boundary layer. Air currents can be made to carry heat by convection in any direction through the use of a fan or a blower as shown in Figure 3-13.

Radiation

Radiation is energy that travels across a space and does not need an intervening medium, such as a solid, gas, or fluid (Figure 3-14). It travels as electromagnetic waves and is very similar to light, radio waves, and x-rays. The radiation waves emanate heat only when the waves can excite molecules; therefore, firefighters can protect buildings from radiated heat by allowing water to run down the face of the building. The water running down the side of the building slows the movement of the molecules on the outside building construction material.

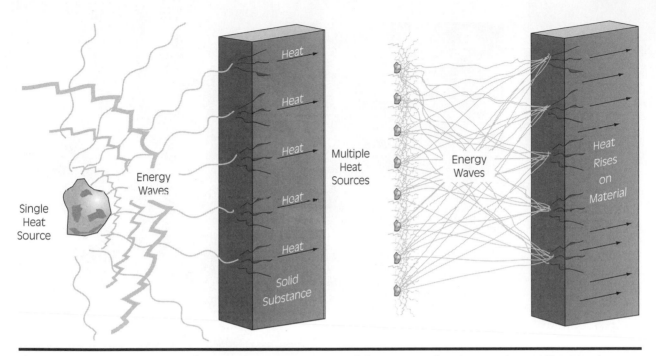

Figure 3-14   *Radiation is energy that travels across a space and does not need an intervening medium, such as a solid or a fluid.*

The difference between radiation and convection heat transfer can be seen through the difference in the behavior of a candle flame. The air that is required for combustion of the fuel vapors is drawn into the flame from the surrounding atmosphere by the process of entrainment. The hot gases rise vertically (lighter than the surrounding air) as a plume that carries most (70% to 90%) of the heat. This heat is released by convection. The remaining heat lost (10% to 30%) from the flames is by radiation. This can be detected by holding one's hand near the side of the flame. The sensation of warmth is caused by radiant heat transfer.

Direct Flame Impingement

In actual fire situations, heat is also transferred by the flames directly impinging upon the materials, thus raising their temperature to the point where combustion occurs (see Figure 3-15). Another way to see the action of direct flame impingement is to think of it as a combination of both convective and radiative processes. The heat is transferred from the plume of hot fire gases by convective heat transfer and radiation until the new fuel pyrolyzes and generates gaseous fuels which are then ignited by the flames.

Figure 3-15 *Flames directly impinging upon the materials transfer the heat, raising their temperature to the point where combustion occurs.*

SUMMARY

The combustion process is defined in various ways, including the type of combustion, the rate of fire growth, the amount of available ventilation, and the type of substance that burns. Some of these descriptions are used in research laboratories for experimental purposes in an effort to better understand and define the combustion process.

In addition, fires are divided into five classifications, organized by the materials that are burning. Using these five classifications, firefighters can identify the most effective extinguishing agent for each of the classes. These classifications are:

- Class A fires (in ordinary combustible materials);
- Class B fires (in flammable liquids);
- Class C fires (in energized electrical equipment);
- Class D fires (in certain metals); and
- Class K fires (in cooking oils).

Firefighters can also use the fire stages and events as a method to provide insight into the condition and extent of a fire. By using this information, they also can control the fire more safely.

The physical and chemical properties of the fuels that feed the fires encountered by firefighters all affect how a fire will burn, how it will spread, and the quickness of its burning rate. This information is vital to firefighters, as it enables them to determine how quickly and safely a fire can be controlled.

KEY TERMS

Backdraft   A sudden, violent reignition of the contents of a closed-container fire that has consumed the oxygen within the space when a new source of oxygen is introduced. The introduction of oxygen results in an immediate smoke explosion.

Carbon monoxide (CO)   The result of the incomplete combustion of a fuel where carbon and a single atom of oxygen are produced.

Cellulosic materials   Materials produced by the chemical modification of cellulose, which is a major constituent of plant life.

Class A fires   Fires involving ordinary combustibles.

Class B fires   Fires involving flammable liquids.

Class C fires   Fires involving energized electrical equipment or wires.

Class D fires   Fires involving combustible metals.

Class K fires   Fires involving cooking oils.

Combustion   A planned and controlled, self-sustaining chemical reaction between a fuel and oxygen with the evolution of heat and light.

Compartmentation   Generally used in high-rise buildings to provide a safe area within the structure where persons can safely be protected from fire.

Conduction   The transfer of heat energy from the hot to the cold side of a medium by means of energy transfer from molecule to adjacent molecule, or atom to atom.

Convection   The movement of heat energy by the agitation of air molecules that reduces the density of molecules, making heated air lighter than cooler air. In a heated enclosed compartment, the heated air rises and pulls in cooler air below the flaming level.

Decay stage   The stage when the fire has consumed all of the available fuel and the temperature begins to decrease as the fire reduces in intensity.

Diffusive flaming   A combustion process where the flames are part of the actual process.

Entrainment   The gathered or captured cooler air that replaces the rising heated air surrounding the point of combustion.

Fire   A rapid, self-sustaining oxidation process that involves heat, light, and smoke in varying quantities. It is often an unplanned or uncontrolled event, as in most cases the fuels are not selected or even known in advance. See also combustion.

Fire tetrahedron   A four-sided model describing the heat, fuel, oxygen, and chemical reaction necessary for combustion.

Fire triangle   An older three-sided model used to describe the heat, fuel, and oxygen necessary for fire. This model is now revised with the introduction of the fire tetrahedron model.

Flame over or rollover   The flames that travel through or across unburned gases in the upper portions of the confined area during the fire's development.

Flaming combustion   An exothermic or heat-producing chemical reaction with flames occurring between a substance and oxygen.

Flashover   A sudden event that occurs when all the contents of a room or enclosed compartment reach their ignition temperature almost simultaneously resulting in an explosive fire.

Free radicals   A molecular fragment possessing at least one unpaired electron. It is chemically active and must react with another free radical to form a compound.

Fully developed stage   The stage of a fire where the maximum generation of heat and consumption of fuel and oxygen occur.

Growth stage   The stage where the fire increases its fuel consumption and heat generation.

Incipient or ignition stage   The point at which the four components of the fire tetrahedron come together and the materials reach their ignition temperature and the fire begins.

Law of latent heat of vaporization   The heat that is absorbed when one gram of liquid is transformed into vapor at the boiling point under one atmosphere of pressure and the result is expressed in BTUs per pound or calories per gram.

Particulates   The unburned products of combustion visible in smoke.

Piloted ignition temperature   The ignition temperature of a liquid fuel. When heated, it will self-ignite.

Pyrolysis   The process of breaking down a solid fuel into gaseous components when heated; also called thermal decomposition.

Radiation   The combined process of emission, transmission, and absorption of heat energy traveling through electromagnetic waves from an area of higher heat energy to an area of lower heat energy. A good example is the transfer of solar heat from the sun to the earth. The heat energy is moved through waves from an area of higher temperature (sun) to an area of lower temperature (earth).

Relative humidity   The ratio of the amount of moisture in a given volume of space to the amount that volume would contain if it were saturated.

Saponification   The process of chemically converting the fatty acid contained in the cooking medium to soap or foam.

Smoldering (glowing) combustion   The absence of flames with the presence of hot materials in the surface where oxygen diffuses into the surface of the fuel.

Spontaneous combustion   An occurrence where a material self-heats to its piloted ignition temperature, then ignites.

Stack effect   **The temperature** difference between the outside temperature of the building and the temperature inside the building. This temperature difference creates air movement inside the building, which may carry heated fire gases to the upper or lower floors.

Stoichiometric   An ideal burning situation or a condition where there is perfect balance between the fuel, oxygen, and end products.

Stoichiometry of reaction   The proportion of fuel to oxygen and the resulting end products.

REVIEW QUESTIONS

1. The new fire tetrahedron model depicts combustion. Explain how each of the four sides of this model depicts the combustion process and how extinguishment is accomplished.

2. Describe the stages of a confined fire and events that may occur if the conditions are right.

3. The term *stoichiometric* is used to describe a burning situation or condition. Explain the meaning of this term.

4. Explain how building construction materials and techniques have changed to increase the amount of heat firefighters find themselves confronted with in a confined space.

5. Describe the four means of heat movement and how firefighters can stop or reduce some of this movement.

REFERENCES

Angle, James et al. (2008). *Firefighting Strategies and Tactics* (2d ed.) (Clifton Park, NY: Delmar Cengage Learning).

Bachtler, Joseph R. and Thomas F. Brennan (eds.). (1995). *Fire Chief's Handbook* (5th ed.) (Saddle Brook, NJ: Fire Engineering, Penwell Publishers).

Delmar Cengage Learning. (2008). *Firefighter's Handbook: Firefighting and Emergency Response* (3d ed.) (Clifton Park, NY: Delmar Cengage Learning).

Friedman, Raymond. (1998). *Principles of Fire Protection, Chemistry and Physics* (Quincy, MA: National Fire Protection Association).

Hawley, Chris. (2008). *Hazardous Materials Incidents* (3d ed.) (Clifton Park, NY: Delmar Cengage Learning).

Klinoff, Robert. (2007). *Introduction to Fire Protection* (3d ed.) (Clifton Park, NY: Delmar Cengage Learning).

Quintiere, James G. (1997). *Principles of Fire Behavior* (Albany, NY: Delmar Cengage Learning).

Chapter

4

EXTINGUISHING AGENTS

Learning Objectives

Upon completion of this chapter, you should be able to:

■ Review and examine the basic components of the fire extinguishment process.

■ Review the five basic classifications of fire and explain the various types of agents used to extinguish or control fires in these five classifications.

■ Examine in detail the variety of agents used for fire extinguishment and explain the application methods for each of these agents.

■ Identify and explain the benefits of using the latest technological advances in fire extinguishing agents such as compressed air foam and ultrafine water mist systems.

INTRODUCTION

Fire is extinguished by eliminating essential elements needed for combustion. For years, the fire service believed that this was accomplished by removing one or more of these factors: fuel, oxygen (air), and heat. As mentioned in the previous chapter, the fire service used a fire triangle or pyramid to depict this process, but other research now shows different processes are involved. Today, we model the fire combustion process using a fire tetrahedron, a four-sided model that includes heat used to release additional vapors from the fuel to make the combustion process continuous. The fire triangle and fire tetrahedron models are reviewed again because a more complete understanding of the fire combustion process helps us to better understand the combustion process and how to be more efficient and effective in extinguishing fires.

Fire professionals use this model daily to extinguish fires, as their task is to select the method that will stop the progress of the fire quickly and safely. By doing so, they reduce the fire, smoke, and water damage as much as possible.

FIRE EXTINGUISHMENT THEORY

Heat
All of the energy, both kinetic and potential, within molecules.

Absolute zero
The temperature at which no movement of molecules occurs.

Temperature
A measure of the average molecular velocity or the degree of intensity of the heat.

Hydrocarbon product
An organic compound, such as benzene or methane, that contains only carbon and hydrogen.

Cellulose material
A complex carbohydrate of plant cell walls used in making paper or rayon.

Heat is a form of energy and can be defined as the molecular motion of the material. The driving force is the effort to maintain equilibrium as heat moves from warm bodies to cooler bodies. There is no such thing as cold, because the situation is relative to another body that is warmer. **Absolute zero** occurs when there is no movement of the molecules. Heat energy above this point indicates that there is some movement of the molecules.

Heat and temperature are not the same measurement. Heat is all of the energy (both kinetic and potential) within the molecules. **Temperature** is a measure of the average molecular velocity (movement) or the heat's degree of intensity. Therefore, heat is the sum of all the material's molecular energy and is measured in calories or British thermal units (BTUs). Firefighters commonly use the number of BTUs per pound of material to assess its relative heat generation during a fire. For example, 1 pound of wood will generate approximately 8,000 BTUs, whereas 1 pound of hydrocarbon products will generate approximately 16,000 BTUs. This means twice as many BTUs will need to be removed or absorbed when extinguishing a fire in a **hydrocarbon product** as would be needed if it were a wood or **cellulose material** fire.

Interruption of the Combustion Process

The combustion process can be terminated or interrupted by removing one or more of the components shown on the fire tetrahedron. When applying the extinguishing agents, firefighters attempt to select the best agent based on the

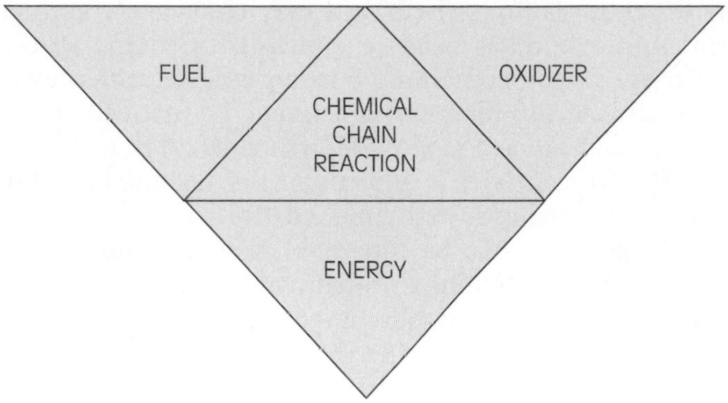

FIRE TETRAHEDRON

FUEL

OXIDIZER

CHEMICAL
CHAIN
REACTION

ENERGY

Figure 4-1   *The fire tetrahedron.*

burning characteristic of the materials. The four main extinguishment processes include:

1. temperature reduction;
2. fuel removal;
3. oxygen removal; and
4. breaking the release of additional fuel from the material.

As illustrated in Figure 4-1, using any of the four extinguishment processes breaks the chain of combustion.

Rekindle
A continuation of a fire, generally after the fire department has declared the fire extinguished and left the scene.

Temperature Reduction/Heat Removal The most common method of fire extinguishment is to use water to cool the temperature of the fuel to below its ignition temperature. The process disables the fuel from releasing combustible vapors. Once the flammable material stops releasing combustible vapors, the combustion process is halted. In structure fires, insufficient cooling can lead to a **rekindle** or reignition after extinguishment.

Solid fuels and liquid fuels with high flash points can be extinguished by cooling. However, flammable vapors may continue to be released. If the fuel temperature is still above the flash point, any source of ignition with sufficient energy will cause a reignition.

Backfire
A fire set to burn the area between the control line and the fire's edge to eliminate fuel in advance of the fire, to change direction of the fire, and/or to slow the fire's progress.

Fuel Removal Removing the fuel source effectively extinguishes fires. The fuel source may be removed by stopping the flow of liquid or gaseous fuel such as shutting down a gas valve or pipeline.

With wildland fires, firefighters remove vegetation in front of the fire, thus eliminating the fuel. This work is accomplished by a bulldozer or hand crews cutting and removing the fuel from the path of the fire. Another removal method used on wildfires is to **backfire** or burn the fuel in front of the main fire thus removing the fuel by fire.

Oxygen Depletion Reducing the amount of oxygen available to the combustion process reduces fire growth and may completely extinguish it. In its simplest form, this method is used to extinguish cooking stove fires when a cover is placed over a pan containing burning food. Another example is where the oxygen content of a room environment can be reduced by flooding the area with an inert gas such as carbon dioxide, which is 1.5 times heavier than air and will replace the oxygen (air) in the room. Because carbon dioxide does not support combustion, the fire is extinguished.

Oxygen can also be separated from the fuel by blanketing the fuel with foam. The foam provides a protective film which, if unbroken, keeps the oxygen from the fuel. Neither of these methods is effective on burning fuels that are **self-oxidizing materials,** meaning that they contain oxygen within themselves and will combust using their internal oxygen.

Chemical Flame Inhibition Some extinguishing agents interrupt the combustion reaction, which is a self-sustaining series of chemical chain reactions. They are needed to keep the fire burning. These extinguishing agents react with the hydrogen atoms or the hydroxyl radicals; however, the exact mechanisms have not been completely determined. When applied, these powders also inhibit the combustion process by absorbing heat.

Extinguishment and Classification of Fires

Classification of fire is based on the type of fuel being burned. Each of the five fire classifications (A, B, C, D, and K) has its own specific requirements for fire extinguishment. For some classes, extinguishment can be accomplished by more than one type of fire extinguishing agent.

THE PROCESS AND AGENTS OF EXTINGUISHMENT

Water has the ability to absorb heat energy more than all other elements with the exception of mercury. Water is relatively inexpensive and is very abundant in most locations. Water is usually the top choice of fire departments, but it may not be the best extinguishment agent for some types of fires.

Water has a nonconforming characteristic in the processes of expansion and contraction. The freezing point of water is 32°F (0°C). When heat is applied to water at this temperature, the water will contract instead of expanding. It will continue to contract until it reaches a temperature of 39°F (3.89°C). Above 39°F (3.89°C), water will expand with regularity as heat is applied until its boiling point of 212°F (100°C) is reached.

In reverse order, if water is cooled it will contract until a temperature of 39°F (3.89°C) is reached then expand until it freezes. When water seeps into

Self-oxidizing materials
Materials containing extra oxygen, which enhances the process of combustion by intensifying the fire.

cracks in rocks or concrete and freezes before it evaporates, the expansion will cause further cracking as the water becomes ice and expands with great force. This same force breaks sprinkler pipes as the water in fire sprinkler piping systems can freeze if not protected.

Of most importance to firefighters is the capability of water to absorb a great amount of heat in the process of converting from a liquid to a gas (steam). This is called the law of latent heat of vaporization, as discussed in Chapter 3.

Latent heat of vaporization is the quantity (or amount) of heat absorbed by a substance when it changes from a liquid to a vapor. At sea level, water boils at 212°F (100°C), and when it reaches 212°F (100°C) it vaporizes or turns into steam. In doing so, the water absorbs heat. For every 1° increase in the temperature in a pound of water, the water will absorb 1 BTU, so when we raise the temperature of the water from 70° to 212°F (21.2° to 100°C), the water will absorb 142 BTUs. There are 8.33 pounds per gallon of water, so 142 BTUs × 8.33 pounds = 1,183 BTUs absorbed from raising the pound of water from 70°F to 212°F; however, continued heating converts the liquid into steam and in this process another 970 BTUs per pound of water is absorbed during the conversion. Because there are 8.33 pounds in a gallon of water (970 BTUs × 8.33 pounds), another 8,080 BTUs is absorbed.

Starting with water in a hose line at 70°F (21.1°C), when applied on a fire, it will be brought up to 212°F (100°C) then converted to steam; and once this conversion occurs, 1 gallon will absorb 1,183 BTUs + 8,080 BTUs = 9,263 BTUs.

In addition to the ability to absorb heat, the high expansion ratio of water to steam can be used to purge smoke-filled areas. The high expansion factor can be seen in the example found in Table 4-1. The table assumes a 90% efficiency of conversion and shows that 50 gallons of water can be expected to expand and occupy a volume of space. Table 4-1 also illustrates the ability of water to purge areas of heated toxic fire gases when converted to steam. The steam also reduces the available oxygen.

Agents to Improve Plain Water

Water is not always the best extinguisher for fires. Gasoline fires or other hydrocarbon liquids with flash points below 100°F (37.8°C) should not be fought with

Latent heat of vaporization
The quantity of heat absorbed by a substance when it changes from a liquid to a vapor.

Table 4-1 *Water expansion properties.*

Temperature (°F/°C)	Cubic Feet of Steam	Volume of Space
212/100	10,000	8 ft × 25 ft × 50 ft
400/204.4	12,000	8 ft × 25 ft × 62.5 ft
800/426.7	17,500	8 ft × 25 ft × 100 ft
1,000/537.8	20,000	8 ft × 25 ft × 100 ft

water. In some cases the fuel simply generates too many BTUs for the water to be effective unless it is applied in copious amounts.

Water is not effective on most metallic dusts and shavings, or in pyrophoric metals. For example, a magnesium shavings fire can result in an explosion as the temperature at which magnesium burns is so hot that it breaks the water down into its basic components, oxygen and hydrogen.

Additives to Improve Water Applications

In an effort to improve the characteristics of water for fire fighting, several chemicals have been developed to increase water's effectiveness when they are combined. Water temperature can be lowered, it can be made thicker, and its surface tension can be reduced to decrease the amount of friction loss. Two of these chemical additives (wet water and slippery water) are used on a regular basis by fire departments.

Wet water
Water with a wetting agent added. The wetting agent reduces the surface tension of the water allowing it to flow and spread better.

Wet Water One additive that reduces the surface tension of water is **wet water.** Reducing the surface tension will allow the water to better penetrate the surface of most other materials. Water can then penetrate further into the materials and result in greater extinguishment capability. Firefighters use wet water when they need to penetrate deeply to reach the seat of a fire in tightly baled materials such as cotton or tightly packed cardboard boxes.

Slippery water
Using polymers, a plasticlike additive to the water, increases the amount of water that could be moved through a hose line.

Slippery water Another additive, **slippery water,** uses polymers, which are plasticlike substances that not only reduce friction loss in the hose but also increase the amount of water that can be moved through a hose line.

Friction loss in fire hoses is caused by water moving through the hose and rubbing against the relatively rough interior of the fire hose. This roughness sets up turbulence in the flowing water, which impedes the smooth flow. More pressure is required to push the water through the hose line. Also, as the amount of flow increases, so does the amount of friction loss, and thus requires the pump to work harder to push the water through the hose and overcome the friction losses caused by rubbing and turbulence of the water against the inside of the hose.

Friction loss
The pressure lost by fluids while moving through pipes, hose lines, or other limited spaces.

Using a plasticlike additive called a polymer not only reduces the friction loss in the hose but also increases the amount of water that could be moved through a hose line. It is possible to get as much water from a 1.5-inch (34-mm) line using a polymer as it is from a 2.5-inch (64-mm) line using plain water. Firefighters should use caution, however, as the amount of nozzle reaction is not reduced, so flowing larger quantities of water will increase the reaction **backpressure, or nozzle reaction pressure.**

Backpressure (nozzle reaction pressure)
The pressure exerted in the opposite direction of the water flowing from a nozzle.

Other Additives to Improve Water Applications

In addition to wet water and slippery water, there are two other products designed to improve water applications, thick water and viscous water.

Thick water
An additive designed to provide an insulating barrier on the surface of a solid fuel.

Viscous water
Thickening agents that are added to water to thicken it so it will cling to the surface of the fuel.

Thick Water Designed to improve water's ability to provide an insulating barrier on the surface of a solid fuel, **thick water** is expensive and reports on its use have not been very positive.

Viscous Water The relative low viscosity of water allows it to run off solid fuel surfaces quickly and limits the ability of the water to blanket a fire by forming a barrier at the surface. In an effort to improve this characteristic, a thickening agent is added. **Viscous water** has several thickening agents added to it. If it is correctly mixed it will:

- cling and adhere to the surface of the fuel by providing a continuous coating which is thicker than untreated water over the fuel surface; and
- project further when discharged from a nozzle and will better resist wind and air currents.

Viscous water has certain negative aspects as well. It will not penetrate fuel as well as water. It also creates a higher friction loss in the fire hose and piping, and increases the water droplet size.

WATER APPLICATION METHODS

Straight Stream Applications

Using a straight bore nozzle should not impact the thermal layer zones (stratification of air and fire gases into layers based on their temperatures) if the nozzle application is directed over the fire, opened quickly, and then shut down. This cools the temperature of the ceiling area over the fire and allows firefighters to enter the room. The quick application should not impact the thermal balance of the room which would raise the temperature in the lower room portions of the room. By maintaining the thermal balance, the area is cool enough for firefighters to be able to stay in the room and make a direct application on the seat of the fire.

The conversion of water to steam is still an effective method of fire extinguishment. It is highly effective on a high flash point liquid fire such as gasoline or some fuel oils. The steam not only absorbs the heat, but also blankets the combustion zone, thus depleting oxygen.

Fog or Spray Applications

In the United States during the years 1900 to 1939, the fire service overlooked the use of spray streams or fog application on fires. The straight bore nozzle was used as the primary application nozzle because it was used daily on structure fires, and the application procedures and methods were fine tuned for maximum results. The nozzle stream was opened and directed at the upper ceiling area directly over the fire to rapidly reduce the overhead heat without

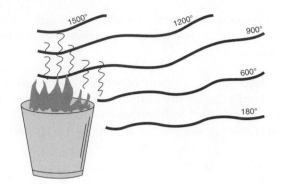

Figure 4-2
Temperature layering.

impacting the thermal layering in the room. After this quick action, the fire was extinguished by directing the straight bore stream of water at the source of the fire. This technique is effective but requires firefighters to get close to the seat of the fire. Nevertheless, once they are near the seat of the fire, they can efficiently apply the water, reducing smoke and water damage to both the contents and the building. Firefighters using the straight bore nozzle did not want to impact the thermal balance of the room environment to maintain a cooler environment, to help them make direct attack to the seat of the fire. Figure 4-2 shows the temperature layering in a confined room. Firefighters identify this as thermal balance.

Figure 4-2 illustrates the movement of the smoke layer, the temperature distribution of hotter air, and the flow of cooler air into the fire room. Note the location of the neutral plane where the thermal balance is reached.

Chief Lloyd Layman, assigned to training for the U.S. Navy, conducted a series of tests on a fog nozzle that was perfected by the Germans. This nozzle divided the water stream into very small drops resulting in more water surface area. The purpose of increasing the surface area of the water droplets is to absorb the heat faster as more water surface area is exposed. This heat quickly converts the water drops into a gas (steam). This conversion process results in more heat being absorbed from the room environment. This method of fire extinguishment is very effective in some circumstances.

When the fog nozzle water pattern is applied to a fire inside a structure, the conversion from water to steam absorbs a great deal of heat; however, it can produce a **thermal imbalance** between the hot smoke layer (located in the upper portions of the ceiling or highest area) and the cool smoke layer zones (in the lower portions) of the room, where firefighters are located. This rapid increase in steam at the lower location can drive firefighters from the room. In Figure 4-3, firefighters have imbalanced the thermal layer, driving the heated upper layer in the lower portions of the room.

In addition to absorbing a great deal of heat, the expansion of water to steam has the ability to purge an area of smoky and noxious gases. Although this

Thermal imbalance
A condition that occurs through turbulent circulation of steam and smoke in the fire area and leads to decreased visibility and uncomfortable conditions.

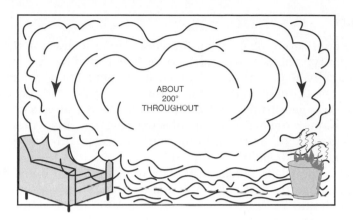

Figure 4-3 *Thermal imbalance.*

is an important factor in reducing the dangerous environment, fires in ordinary combustible materials are normally extinguished by the absorption of heat, not by the smothering effect created by the steam.

Both straight bore nozzles and fog nozzles have a specific use in fire fighting procedures. It is up to the professional firefighter to size up the situation and select the best choice for the fire problem presented. (See the references for additional reading on this issue.)

Water Mist Systems

Over the years, the idea of using an ultrafine water mist for fire suppression purposes has been generally disregarded based on the belief that such exceptionally small droplets would not have enough energy to reach the fire and would vaporize before they would contact the fire area.

However, new research and technological improvements in nozzles and pressure systems have demonstrated that efficient fire suppression capability is possible but depends on mist droplet size, mist stability, the transportation behavior of the mist, unobstructed nozzles, and the efficiency and rate of droplet vaporization. Although additional development is needed, it appears that water mist can replace the use of the ozone-depleting chemical agent Halon 1301, which is now banned, and may replace the clean gas chemical agents such as new clean agent heptafluoropropane (HFC-227 ea.).

Firefighters can expect to encounter high-pressure water mist and clean agent systems in any location where water may not be appropriate for fire extinguishment, such as areas with computers, servers, or specialized electronic equipment.

Firefighters need to be aware that applying water mist provides additional hazards for structural firefighters. First, the almost complete conversion of the water mist to steam drastically reduces the visibility; and second, if the water is totally converted to steam, it will increase the enclosure temperature.

■ **Note**

Foam is created by using a process of agitation with air injection or aspiration to mix three ingredients together: water, foam concentrate, and air. Water is the primary component of foam ranging from 94% to 99% depending upon the ratio of water to foam concentrate.

Foams

The application of foam on fires has a long history of success, as long as the proper type of foam was selected for the class of the fire and the circumstances. Originally, foam was made by mixing two separate powders with water in a generator that combined the foam products and generated carbon dioxide, which became trapped inside the foam bubbles. The result was a thick blanket of foam containing carbon dioxide bubbles working to insulate the foam blanket as it was spread over the burning liquid surface. Later, this two-powder foam was replaced with single-powder foam.

In the 1940s, mechanical foam was introduced to the fire service. Prepared from animal products, it is now commonly known as protein foam. It was mixed with air and had a very high expansion ratio. It was a free-flowing material and because of its ease of handling, protein foam became popular. However, it deteriorates rapidly and needs to be used quickly after purchase as it has a very short shelf life.

A few years later, surfactant or detergent foam became available. It was more compatible with dry chemicals which destroyed protein foam and it had a much longer shelf life than protein foam. The surfactant or detergent foam could better penetrate a porous solid fuel source as it reduced the surface tension of the water and, as a result, came to be known as wet water or a wetting agent.

Foam Classifications

Class A foam
Foam intended for use on a Class A fire.

Class B foam
Foam intended for use on a Class B fire.

Compressed air foam system (CAFS)
A technology developed for foam-producing systems which are generally installed on the fire engine or special unit.

Foams are classified as Class A, Class B, or special foams. **Class A foams** are designed to extinguish Class A fires, or fires in combustible solids such as wood, fabrics, paper, or organic materials. **Class B foam** is intended for Class B fires such as oil, greases, and other hydrocarbon products. There are a number of specially formulated foams for other products firefighters may encounter. One example is alcohol-type foaming agents. **Compressed air foam systems (CAFSs)** are becoming a very popular option for fire departments to assist in a wide variety of fire classification. Students are encouraged to research further to find other foaming agents that may be better for specialized fires.

Foams are also defined by their expansion ratio into three categories. The first is low expansion foam, where the bubble expansion ratio is small (less than 20:1) and the bubble contains a high percentage of water. The other two are medium and high expansion foam, where the expansion ratios are greater than 20:1 and some varieties up to 1,000:1. Medium expansion foam has an expansion ratio between 20 and 100, while high expansion foam has an expansion ratio above 200. At these expansion ratios, the bubble water content is low and as a result the bubble is relatively light. The foam bubbles are made by mixing a foam concentrate with water and a foam solution. The foam solution is then mechanically agitated to form bubbles.

Foams are also defined by their effectiveness on hydrocarbon fuels, water-miscible fuels, or both. Low expansion foams generally are applied to fires involving flammable or combustible liquids in storage tanks. These tank systems discharge foam bubbles over the liquid surface to provide a cooling, smothering blanket that progressively covers the liquid surface and extinguishes the fire. This foam blanket can prevent vapor production for some time. Low and medium expansion foams provide a low viscosity, are mobile, and are able to cover large areas quickly. High expansion foams are for enclosed spaces such as aircraft hangers where quick deployment is necessary.

In the early 1960s, research by the U.S. Navy led to the development of a film-forming foam "light water," so-called because it leaves a thin film of water on the top of the liquid fuel source. The extremely thin (0.001 inch) film prevents the formation of fuel vapor. It is compatible with all dry chemicals and has led to a dual-agent fire attack system which is now being encouraged by Federal Aviation Administration (FAA) regulations for airport fire protection. This type of foam is known as **aqueous film-forming foam (AFFF)**.

Aqueous film-forming foam (AFFF)
The type of foam created by a combination of water and perfluorocarboxylic acid.

AFFF Concentrate

AFFF concentrate can be applied with less sophisticated foaming devices such as regular fire department spray nozzles and sprinklers (unlike most other foaming agents). This foam concentrate is available for proportioning to a final concentration of 1%, 3%, or 6% by volume with either freshwater or seawater. The strength of the film surface and its fluidity (ability to flow) on kerosene makes it a successful extinguishing agent for jet aircraft fuel. However, this foam loses water content rather rapidly (drain time) and may provide less burn-back resistance compared to other protein-based foams. The **drain time** is the amount of time required to lose the water from the mixture.

Drain time
The time it takes for the foam to break down as the water separates (drains) from the aerated foam bubbles when they break.

Application of Class A Foams

Class A foams consist of a mixture of water, foam concentrate, and air. The foam mixture can be made wetter or drier by adjusting the ratio of the foam solution to the amount of water. The size of the air bubbles is determined by the adjustment of the amount of air pressure as the foam can be made more dense with less air and lighter with more air. The size of the bubbles also depends on materials burning and if the foam (water) needs to penetrate the materials or cover the surface of the material. See Figure 4-4 for the various application modes of Class A foam.

Wet foam Characterized by smaller bubbles, wet foam has less expansion because less air has been introduced, and it has fast drain times. The drain time is the amount of time for the water to drain away from the foam solution. Wet foams are generally good for initial fire suppression, overhaul, and penetration into deep-seated fires.

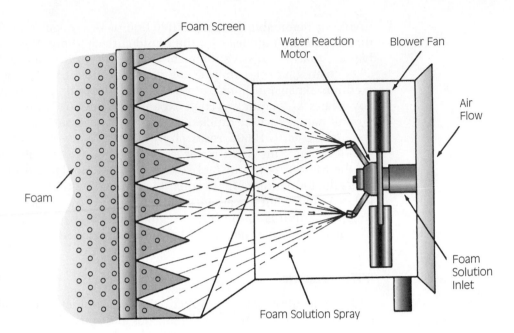

Figure 4-4 *Foam expansion ratio and drain time.*

Dry foam Dry foam has a high expansion ratio and is the consistency of shaving cream. The dry foam is very fluffy and consists mainly of air. Dry foams have slow drain times and hold their shape for a long period of time. Dry foam is especially good for exposure protection because of its ability to adhere to vertical surfaces for long periods of time.

Fluid foam Fluid foam has the consistency of watery shaving cream. Fluid foam tends to have medium to smaller bubbles and moderate drain times. Fluid foam works well for direct attack, exposure protection, and mop-up operations.

High-pressure compressors mix the foam solution with air and then deliver it through the fire hose at 100 psi air pressure. This process adds the needed bubble agitation to the finished foam product. Figure 4-5 illustrates the process of making high expansion foam.

FIRE EXTINGUSIHING CHEMICALS AND OTHER AGENTS

Researchers have found that water is not always the best extinguishing agent for all fires. Therefore, a number of specially formulated dry chemicals, carbon dioxide, and halogenated agents have been developed to fill in where water fails to do the job.

Figure 4-5 *Making high expansion foam.*

Dry Chemicals

Various dry chemical materials are commonly used as extinguishing agents. The principle chemical used for dry chemical agents is sodium bicarbonate, potassium bicarbonate, potassium chloride, and urea-potassium bicarbonate. Using these agents with specific additives can improve their storage, flow, and water repellency.

They work exceptionally well on flammable liquid fires such as grease, some oils, and gasoline (Class B fires). Some dry chemicals are designed to work on metal fires (Class D fires).

Systems using dry chemicals are primarily used to protect flammable or combustible liquid storage locations. In these systems, the dry chemical agent can be discharged through fixed piping, supplying nozzles or hose lines, or they can be designed to protect hazardous operations such as a dip tank or other process hazard.

Discharge nozzles also can be arranged to discharge the extinguishing agent on a burning surface such as cooking oil or grease (Class K fires), grease filters, and grease exhaust ducts. See Figure 4-6 for an example of a stored pressure dry chemical extinguisher.

In these systems, the agent is discharged under high pressure (approximately 350 psi) by a gas expellant, generally carbon dioxide or nitrogen. The dry chemical forms a blanket over the top of the fire and extinguishes by excluding the oxygen. Dry chemical extinguishers are also available in a smaller, portable

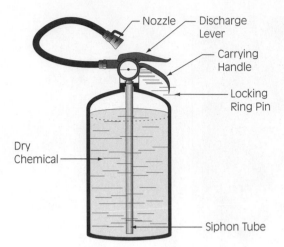

Nozzle
Discharge Lever
Carrying Handle
Locking Ring Pin
Dry Chemical
Siphon Tube

Figure 4-6 *Stored pressure dry chemical extinguisher.*

container or a larger wheeled model and some are designed to be shoveled by hand onto the burning surface.

Application of Dry Chemicals

Two common arrangements are used for dry chemical expellant gas (stored pressure), or the dry chemical can be stored in a separate pressure container that is normally under atmospheric pressure and is connected to a pressurized cylinder (cartridge operated). If the dry chemical agent is stored in a separate container, a rupture disc is installed in the piping between the pressure tank and the agent storage tank. Activation of the system can be manual, automatic, or both. These systems usually have a manual release mechanism for the dry chemical storage cylinder that is located far enough from the hazard so the operator would not be burned while attempting to actuate the system.

The location of the discharge nozzles is important, as they need to be located in the center above the cooking surface. One problem is that most cooking equipment has wheels installed so it can be moved for cleaning, and after cleaning it must be returned to the same location in order for the system to be effective in the event of a fire. Another concern is that the nozzles are often attached to piping that extends down from the hood as shown in Figure 4-7.

Carbon Dioxide

Carbon dioxide fire extinguishers consist of high-pressure cylinders or low-pressure tanks containing carbon dioxide (CO_2) under pressure. These cylinders can be discharged through a special hose and nozzle or connected to fixed piping with nozzles. Both systems work by flooding the area with carbon dioxide that is heavier than air (1.5 times), thus excluding the supply of oxygen and suffocating the fire.

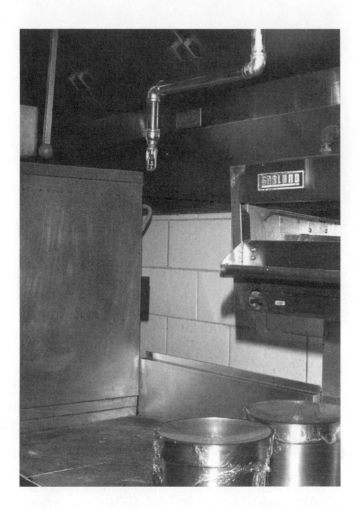

Figure 4-7 *In dry chemical systems protecting kitchen equipment, nozzles are often attached to piping that extends down from the hood.*

Carbon dioxide is a colorless, odorless, electrically nonconducive gas and is noncorrosive to most metals. Because of these characteristics, it is frequently used for protection of electrical equipment and is also effective on most flammable liquid fires. Where the protected hazard is enclosed in a room, total flooding systems are used. Since total flooding systems extinguish the fire by smothering, the entire volume of space within the room being protected must be completely enclosed to make the system effective. Some of the systems have doors that automatically close when the system is activated, therefore firefighters must wear a self-contained breathing apparatus (SCBA) at all times when entering the room.

Application of Carbon Dioxide Carbon dioxide has been obtainable as a commercial product for years. It is known as "dry ice" and is used as a refrigerant or in its

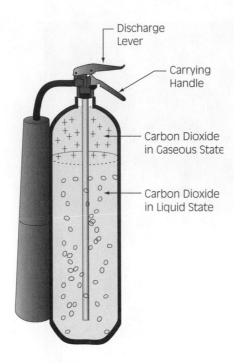

Discharge
Lever

Carrying
Handle

Carbon Dioxide
in Gaseous State

Carbon Dioxide
in Liquid State

Figure 4-8 *Low-pressure carbon dioxide extinguishing system.*

gaseous form for carbonating beverages. Carbon dioxide extinguishes fires by smothering (reducing the oxygen content below 15%). For people exposed to this reduced oxygen environment, such a limitation of oxygen can be lethal.

When carbon dioxide discharges, it creates a refrigerating effect, which can condense water vapor in the area, producing a fog that reduces visibility. In addition, the movement of carbon dioxide through the extinguisher nozzle creates a shrill noise, which can be frightening to those not aware of the sound of escaping gas. The movement of carbon dioxide under pressure through the system while changing state from a liquid to a gas creates static electricity. The carbon dioxide extinguishers have plastic or wooden handles to protect against static electricity. See Figure 4-8 for an example of a carbon dioxide extinguisher.

For enclosed rooms, a predischarge warning alarm is required for total flooding systems in rooms that may be occupied. The alarm warns all persons to evacuate prior to carbon dioxide discharge.

Halogenated Agents

The development and use of halogenated fire extinguishing agents has slowly evolved over the years with improvements and refinements of the chemical compounds. The search for a more efficient and less damaging extinguishing agent than water was needed for locations where water-sensitive electrical equipment or water reactive materials were present.

Some halogenated fire extinguishing agents such as carbon tetrachloride were efficient extinguishing agents but deadly to humans when operated in a closed compartment. After several deaths, the search began to find an extinguishing agent that was efficient in extinguishing fires in sensitive electrical equipment but not toxic to humans.

After years of development, a product known as Halon 1301 emerged as an effective extinguishing agent that was not a serious toxic threat to humans. The fire service community embraced the use of Halon 1301 and many extinguishing systems were installed, particularly where total flooding could be achieved.

In the 1980s and early 1990s, research indicated that the halogenated extinguishing agents seriously depleted the ozone layer surrounding the earth. This depletion allows harmful rays from the sun to enter into the earth's atmosphere.

In 1994, production of certain halons used for extinguishing purposes was banned in certain countries. Halogenated systems at the time of the banning were allowed to be phased out when their in-service life required recharging the system. While certain halons are efficient and effective fire extinguishing agents and are of little threat to humans as far as toxicity, they do present a serious threat to the environment. Their use should be limited to only those locations where release for a fire is absolutely necessary and the halon can be recaptured and returned to the confining equipment. Once used, the system must be replaced with extinguishing systems not impacting the environment. The two banned halons are Halon 1211, a liquid form, and Halon 1301, a gaseous form. Both Halon 1211 and Halon 1301 were used to protect electronic equipment as they left no residue. As a result of the banning, new "environmentally clean" agents have been developed.

Alternative (Clean) Halogenated Fire Extinguishing Systems

A clean fire extinguishing system is employed for protection against fire for electrical and electronic telecommunication facilities, flammable and combustible liquids and gases, and other high-value properties where water damage may be of concern. Some of the systems likely to be encountered by firefighters are listed in Figure 4-9.

Figure 4-9
Alternative halogenated extinguishing systems.

- IG-55 (nitrogen and argon)
- IG-541 (nitrogen, argon, and carbon dioxide)
- IG-01 (argon)
- HFC-236a (hexafluoropropane)
- HFC-23 (trifluoromethane)
- HFC-227ea (heptafluoromethane)
- HFC 125 (pentafluoropropane)
- HCFC Blend A, HCFC-124 (chlorotetrafluoroethane)
- FC-3-1-10 (perfluorobutane)

These clean agents are not effective against fires in certain chemicals or mixtures of chemicals such as cellulose nitrate and gunpowder that achieve rapid oxidation in the absence of air. They are not useful in protecting against fires in reactive metals, such as lithium, sodium, potassium, magnesium, titanium, zirconium, uranium, and plutonium or in metal hydrides.

Furthermore, clean agents do not provide satisfactory protection for chemicals undergoing auto-thermal decomposition such as certain organic peroxides and hydrazine. These liquefied agents might cause electrostatic charging during their discharge, creating an electric arc that could initiate an explosion. A static spark may occur because as the materials move through the piping it creates friction against the walls and will build up an "electrical charge." Grounding both the sending vessel and receiving vessel diminishes the charge, which prevents an electrical charge from developing.

These systems are designed primarily for areas where persons are prohibited. Where spaces are occupied, a time delay is essential to permit evacuation before discharge of the extinguishing agent. All of the systems are designed for discharge into an enclosure that can maintain the gas concentration. Monthly inspection of the enclosure is needed to ensure that openings have not been made in the enclosure without being properly sealed. Air-handling systems serving the protected area need to be shut down and doors and dampers closed, prior to agent discharge.

Clean Agent Application Methods   Triggering the detectors or a manual release starts an alarm sequence in the control panel. The control panel can be designed to sound alarms, shut down power, operate the building fire alarm system, and perform other operations including operating the control head to release the extinguishing agent.

The control head sits on top of the discharge control valve. Upon operation, the control head provides a gas escape route into the atmosphere through the discharge control valve. The escape of gas creates an imbalance of pressure in the discharge control valve, causing the gas in the cylinder to raise a piston in the valve; this in turn uncovers the discharge port, allowing the agent to flow out of the cylinder.

Piping and nozzles are installed on the discharge port. Nozzles are designed to distribute the extinguishing agent at a controlled rate, smoothly and evenly in fan-shaped patterns. The agent is discharged at a right angle to the nozzle in 180-degree or 360-degree patterns. Nozzles should be covered with blow-off caps to prevent entry of foreign material into the pipes or nozzles.

SPECIAL EXTINGUISHMENT SITUATIONS

Many metals and some chemicals are incompatible with water, dry powder, and carbon dioxide extinguishing agents when burning. In some cases, these metals and chemicals react violently when these incompatible agents are used.

Therefore, specialized extinguishing agents are required for a number of them. Firefighters need to be familiar with the characteristics as well as the need for special extinguishing agents for each of these materials as well as how to apply these special extinguishing agents.

Combustible Metal Fires

For fires involving combustible metals, water is generally an improper extinguishing agent as a number of burning metals react violently upon water application. On some of these fires the water is separated into hydrogen and oxygen which further increases the combustion process.

Furthermore, some molten metal fires are very difficult to extinguish because violent steam explosions can occur if the water enters a pool of molten metal. The molten metal pool also will retain the heat for long periods of time and in doing so may result in further fire extension.

Over the years, many specialized fire extinguishing materials have been developed for specific metals. It is imperative that firefighters become familiar with the occupancies utilizing or storing these metals and check with the prevention bureau to ensure that the correct extinguishing agent is required to be on the premises. Table 4-2 contains a brief listing of some of these extinguishing agents.

Table 4-2 *Extinguishing materials for some metal fires.*

Extinguishing Material	Main Ingredients	Metal Can Be Applied On
A liquid that can be applied		
TBM	Trimethoxyboroxine	Mg, Zi, Ti, Na, K
Gases that can be applied		
Helium	BF_3	Any metal
Argon	Ar	Any metal
Nitrogen	N_2	Any metal
Other materials that can be applied		
Met-L-X	$NaCl + Ca_3(PO_4)_2$	Na
Foundry flux	Mixed chlorides + fluorides	Mg
Lith-X	Graphite + additives	Li, Mg, Zi, Ti, Al
Dry sand	SiO_2	Various metals
Sodium chloride	NaCl	Na, K
Soda ash	Na_2CO_3	Na, K
Lithium chloride	LiCl	Li
Zirconium silicate	$ZrSiO_4$	Li

Chemical Fires

In addition to metals, water is not compatible with certain inorganic chemicals. For example, calcium carbide reacts with water to form acetylene, a highly flammable gas with a wide explosive range (12% to 74%). Firefighters may encounter calcium carbide in high school or college chemistry laboratories or chemical plants where production of acetylene gas used for welding is occurring. Other alkali metals such as lithium hydride and sodium hydride react with water to form hydrogen gas, another highly flammable gas with a wider (than acetylene) explosive range of 4% to 81%. Both lithium hydride and sodium hydride can be found in manufacturing plants where porcelain, ceramic castings, and some greases are produced.

The peroxides of sodium, potassium, barium, and strontium react exothermically, releasing great amounts of heat energy when subjected to water. Some cyanide salts such as potassium cyanide react with acidified water (water with acid) to form a highly toxic (and deadly) hydrogen cyanide. Firefighters can encounter these deadly gases in leather tanning plants and wool dying processes.

The problem with applying water to fires involving toxic chemicals such as pesticides is the runoff of contaminated water, which may impact groundwater sources. In every case where a chemical fire is encountered, it is imperative that firefighters make provisions to have persons who can provide technical expertise and advice respond to the incident.

Pressurized Gas Fires

These fires are somewhat difficult to extinguish because the gas is under pressure and a damaged tank, fitting, or valve will provide a continuous supply of fuel. Where there is no fire, immediate evacuation of the area is required. In those cases where the gas supply can be shut off or the tank plugged, firefighters need to make sure actions taken are under the protection of hose lines.

Where the leaking fuel is burning or the tank is being impinged by fire from another burning object, firefighters need to make sure that once the gas supply is shut off, any heated objects are cooled or other sources of ignition are extinguished to prevent reignition. If the damaged tank cannot be plugged or the valve repaired, firefighters must work to keep the tank cool to a point that the pressure relief valve can safely release the gas pressure to the atmosphere.

If it is not feasible or possible to cool the tank where fire is impinging on the tank, unmanned master streams should be put in place to cool the tank and all personnel should evacuate the area because a rapid increase in tank pressure may cause the tank to BLEVE. Fires in pressurized flammable tanks are difficult and dangerous. In Figure 4-10, the events leading up to a BLEVE have been diagrammed to show the conditions of the tank prior to the BLEVE.

■ **Note**
Firefighters should carefully develop pre-incident plans for areas in their jurisdiction where these tanks are stored and where these pressurized flammable liquid tanks travel. Then review the numerous reports detailing actions of firefighters at BLEVEs that have resulted in fatalities.

■ **Note**
Firefighters should avoid attacking the tank fire from either end of the tank, as these are the locations where the tank is welded together. They are the two weakest points of the tank.

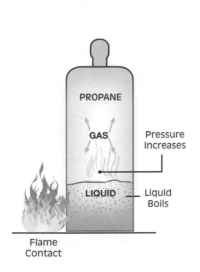

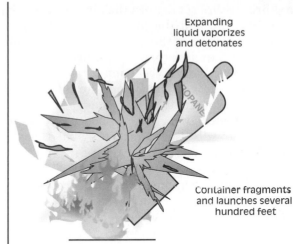

Figure 4-10
Diagrams of BLEVE events.

SUMMARY

It is important to understand the essentials of fire extinguishing agents. Fires can be extinguished by cooling, depleting the oxygen, removing the fuel, or breaking the chain of combustion by blocking the release of free radicals from the fuel.

Although water is abundant in most areas, and thought to be an inexpensive extinguishing agent when all of the costs of delivering water to the fire scene are added up, we find that the costs are fairly high. Nevertheless, water is still inexpensive and available for the majority of fires encountered by firefighters. Water can also be made a more effective extinguishing agent by making it slippery, reducing its surface tension, and making it thicker.

The other classifications of fires require the professional firefighter to select the proper extinguishing agent for application. Many of these agents may be effective on several types of fire, but most are best suited for one type of fire class. These other agents—carbon dioxide, dry chemical, and foam—are effective on certain materials and under certain situations. The recent findings of the ozone-depleting characteristics of halon extinguishing agents has led to a newer clean version (HFC-227 ea), but research is now being conducted on ultrafine droplets of water to provide a clean extinguishing method for water-sensitive locations.

KEY TERMS

Absolute zero The temperature at which no movement of molecules occurs.

Aqueous film-forming foam (AFFF) The type of foam created by a combination of water and perfluorocarboxylic acid. When used, the surface tension of the fuel exceeds the aggregate surface tensions of the film. Oxygen is then excluded from mixing with the fuel vapors.

Backfire   A fire set to burn the area between the control line and the fire's edge to eliminate fuel in advance of the fire, to change direction of the fire, and/or to slow the fire's progress.

Backpressure (nozzle reaction pressure)   The pressure exerted in the opposite direction of the water flowing from a nozzle.

Cellulose material   A complex carbohydrate of plant cell walls used in making paper or rayon.

Class A foam   Foam intended for use on a Class A fire.

Class B foam   Foam intended for use on a Class B fire.

Compressed air foam system (CAFS)   A technology developed for foam-producing systems which are generally installed on the fire engine or special unit. The amount of wetness or dryness of the foam can be set. Once programmed, the special unit automatically regulates the amount of water, foam concentrate, and air pressure. It manufactures a uniform mixture of foam bubbles.

Drain time   The time it takes for the foam to break down as the water separates (drains) from the aerated foam bubbles when they break.

Friction loss   The pressure lost by fluids while moving through pipes, hose lines, or other limited spaces.

Heat   All of the energy, both kinetic and potential, within molecules.

Hydrocarbon product   An organic compound, such as benzene or methane, that contains only carbon and hydrogen.

Latent heat of vaporization   The quantity of heat absorbed by a substance when it changes from a liquid to a vapor.

Rekindle   A continuation of a fire, generally after the fire department has declared the fire extinguished and left the scene.

Self-oxidizing materials   Materials containing extra oxygen, which enhances the process of combustion by intensifying the fire.

Slippery water   Using polymers, a plasticlike additive to the water, it was found that the polymers not only reduced the friction loss in the hose but also increased the amount of water that could be moved through a hose line.

Temperature   A measure of the average molecular velocity or the degree of intensity of the heat.

Thermal imbalance   A condition that occurs through turbulent circulation of steam and smoke in the fire area and leads to decreased visibility and uncomfortable conditions.

Thick water   An additive designed to provide an insulating barrier on the surface of a solid fuel.

Viscous water   Thickening agents that are added to water to thicken it so it will cling to the surface of the fuel. Because it is thicker, it also provides a more continuous coating over the fuel surface and absorbs more heat.

Wet water   Water with a wetting agent added. The wetting agent reduces the surface tension of the water allowing it to flow and spread better. The improved flowing characteristic allows the water to penetrate tightly packed goods better than water not treated.

REVIEW QUESTIONS

1. Saponification is a process that quickly converts the burning substance to a noncombustible soap. Once the process has made the conversion, it extinguishes the fire by what action?

2. What reaction would you expect of magnesium shavings under fire conditions when water is applied?

3. Explain the concept of fire extinguishment used by water mist systems. What are the problems with water mist systems and what are the benefits of their use?

REFERENCES

Bevelacqua, Armando S. (2001). *Hazardous Materials Chemistry* (Clifton Park, NY: Delmar Cengage Learning).

Delmar Cengage Learning. (2008). *Firefighter's Handbook: Firefighting and Emergency Response* (3d ed.) (Clifton Park, NY: Delmar Cengage Learning).

Gagnon, Robert. (2008). *Design of Special Hazard & Fire Alarm Systems* (2d ed.) (Clifton Park, NY: Delmar Cengage Learning).

———. (2003). *Fire Protection Systems and Equipment* (Clifton Park, NY: Delmar Cengage Learning).

Klinoff, Robert. (2007). *Introduction to Fire Protection* (3d ed.) (Albany, NY: Delmar Cengage Learning).

Lowe, Joseph D. (2001). *Wildland Firefighting Practices* (Albany, NY: Delmar Cengage Learning).

Quintiere, James G. (1998). *Principles of Fire Behavior* (Albany, NY: Delmar Cengage Learning).

Sturtevant, Thomas B. (2005). *Introduction to Fire Pump Operations* (2d ed.) (Clifton Park, NY: Delmar Cengage Learning).

Chapter

5

FOUNDATIONS OF FIRE FIGHTING TACTICS AND STRATEGIES

Learning Objectives

Upon completion of this chapter, you should be able to:

- Describe the process of developing the fire fighting strategy and tactics involved in planning, locating, confining, extinguishing, and overhauling fires in buildings and other special fire situations.

- Discuss the methods used for the determination of the proper fire operating mode: offensive, transitional, defensive, or nonattack mode.

- Define the term *size-up* and explain the steps and factors involved in making a size-up.

- Review the fire strategy and tactics used by firefighters and apply the fire behavior characteristics discussed in the text to the situations reviewed.

- Describe difficult fire situations encountered by firefighters and the strategies and tactics they should use to resolve these situations.

INTRODUCTION

This chapter concentrates on structure fires and fire behavior patterns most likely to be encountered. The starting point is a description of the decision-making process used by firefighters to determine the mission, strategy, and tactics to use on emergency incidents. The overall mission for fire incidents has three major components: finding the location of the fire, confining the fire, and extinguishing the fire. These components serve as the foundation of concern for setting the overall incident strategy as well as the priority of the activities to complete the plan. Once these components are considered, the incident commander can then begin to look at the specific steps in the incident size-up.

Size-up

A method used by firefighters to identify the problem(s) presented by the incident.

Size-up is a method used by firefighters to identify the problem(s) presented by the incident. Once the problems are identified, size-up procedures are used to determine the best overall strategy to be used to solve the problem. The tactics are those actions needed to complete the strategy or plan.

Once the size-up is completed and the problem(s) identified, the strategy is determined. The tactics to resolve the problem(s) are then implemented.

The latter part of this chapter examines how to apply fire behavior tactics to specific occupancies such as special fire situations or areas proven to be difficult or dangerous. The fire behavior principles discussed in the earlier chapters of this text should always be kept in mind and applied for a better understanding of how a fire will behave. This chapter begins with a description of the mission or purpose, then discusses the framework for developing the strategy and tactics, and lastly discusses specific tactical actions.

DEVELOPMENT OF STRATEGY AND TACTICS

The overall mission for firefighters responding to fire incidents is:

1. locate the fire;
2. confine the fire; and
3. extinguish the fire.

These three areas are accomplished quickly and safely when the correct strategy and tactics are developed for the situation. Figure 5-1 illustrates where arriving units are locating, confining, and extinguishing a fire.

Decision-making model

A five-step process used to solve problems.

Developing the strategy and tactics for an emergency incident is a five-step process. The first step is the identification of the problem(s) posed by the incident. To start developing the overall strategy on emergency incidents, the **decision-making model** is used. The five steps of the decision-making model, shown in Figure 5-2, are simple.

Figure 5-1 *On fire incidents, the overall mission is to locate, confine, and extinguish the fire.*

The Decision-Making Model

1. Identify the problem.
2. Identify solutions to solve the problem.
3. Select the best solutions(s) based on the situation or circumstances.
4. Implement the solution.
5. Reexamine the solution to see if it is working. If not, start over.

Figure 5-2 *The five steps in the decision-making model.*

The first step is to quickly summarize and carefully define the problem(s) presented. This is done by using the size-up acronym representing components to be examined, to see if it poses a problem that may be of concern on this incident. Once the exact nature of the problem is identified, the second action is to list possible alternatives. Third, select the best alternative to resolve the problem based on the information available. Once the best alternative has been selected, the fourth action is to implement the solution. After the strategy is put into place, reevaluate to see if the expected outcomes occur. If not, reexamine the problem and the solutions, select the best alternate solution, and begin again.

The important point to remember is that the development of the strategic plan is based on gathering as many facts as possible by using the decision-making process to identify the problem. If used on every response, this decision-making process becomes natural. It will also provide a method for which to consider the problems faced on emergency incidents.

Attack Modes

An important aspect of the size-up process is deciding the best mode of attack on a fire. The attack modes are the **offensive mode,** the **transitional mode,** the **defensive mode,** and the **nonattack mode.** The nonattack mode is sometimes referred to as a **passive approach.**

Offensive Attack Mode The most effective means of fire extinguishment in many situations is the offensive or direct attack mode. Firefighters using the direct attack mode enter the building with a straight or solid hose stream to deliver water directly to the fire base. The water cools the fuel below its ignition temperature. The application is in short bursts and in a controlled manner so that water is conserved. A fire in an occupied building is one example of a situation where the offensive mode may be called for if safety procedures and backup crews are in place.

If the decision has been made to conduct an offensive attack, the rule of thumb used by fire officers is, if after twenty minutes of an aggressive, offensive attack the fire is still gaining headway, then changing to a defensive or transitional mode should be considered.

Transitional Mode The transitional mode redirects the strategy because of changing circumstances. It is a phase that is passed through, as the operations are shifted from one mode to another. Care needs to be taken because hose lines protecting firefighters' positions are being moved or can be accidentally shut off, leaving them exposed to fire, heat, and smoke.

Defensive Attack Mode The need to confine the fire inside a structure is the purpose of a defensive attack mode. During a fire attack, the defensive attack mode is used in situations where the temperature is increasing and it appears that the room is ready to go into a flashover event. A short burst of water is aimed at the ceiling to cool the superheated gases in the upper levels of the room. This action can delay flashover long enough for firefighters to apply water using the direct attack method by directing water to the seat of the fire or to make a safe exit from the environment. Water applied to the ceiling area should only be enough to cool it down slightly; too much cooling will impact the balance or the thermal layering in the confined space. See the discussion in Chapter 4 on the effects of an unbalanced thermal layer.

The defensive mode is also used when a building is fully enveloped in fire and is threatening structures nearby. In this attack mode, efforts are made to protect exposed properties.

Nonattack Mode Some emergency incidents present with conditions posing an undue threat to firefighters' lives. One example is a potential BLEVE, which should be considered at all incidents where closed containers of liquids (such as drums, cans, and tank cars) are exposed to fire. The nonattack mode also is

Offensive mode
Fire fighting operations that make a direct attack on a fire for purposes of control and extinguishment.

Transitional mode
The critical process of shifting from the offensive to the defensive mode or from the defensive to the offensive mode.

Defensive mode
Generally a fire strategy which is conducted on the exterior of the building to protect the adjacent buildings from fire spreading.

Nonattack mode (passive approach)
Under certain circumstances, a fire attack may be too dangerous and the incident command will choose to let the fire burn out without an attack.

recognized as an important tactic when faced with fires involving poisons, pesticides, or paint warehouses where runoff water from hose streams could spread toxic materials into groundwater. One example is a paint factory fire where water runoff into the city groundwater reservoir could seriously impact the city water supply. In this situation, allowing the fire to continue to burn is considered the safest, least costly, and fastest means to resolve the problem.

The decision to use the nonattack mode cannot be made without careful consideration and by making every effort to include the owner in the decision-making process. Generally, if the building owner can be contacted and is given a complete explanation of the situation including an analysis of the risks, the owner will better understand the strategic decisions of the fire department.

SIZE-UP AT THE INCIDENT SCENE

RECEO-VS
An acronym used when developing the strategy and tactics on the fire ground.

To improve the size-up process, the fire service has developed a number of acronyms to assist in developing the size-up. The first one presented by the National Fire Academy staff is **RECEO-VS.** Each letter represents one of the seven components used to develop a strategy. The first five are in priority order; the last two may be used at any point to support the first five. RECEO-VS stands for:

> **R**escue
>
> **E**xposures
>
> **C**onfinement
>
> **E**xtinguishment
>
> **O**verhaul
>
> **V**entilation
>
> **S**alvage

This acronym is quicker to use than the second acronym, but is not as comprehensive. The acronym COAL T'WAS WEALTHS contains a mental checklist of fifteen items. It is a reminder of the various components that need to be considered when looking for the problem(s) at an incident. Each of the components helps the firefighter to determine the needs for the specific emergency at hand (see Figure 5-3).

Each of the following components is described to provide firefighters with a detailed mental checklist of items to be considered during the development of the strategy.

Construction

The type of construction includes the building components, materials, and the extent of their fire-resistive abilities. The National Fire Protection Association (NFPA) has divided the construction of buildings into five major classifications. The local building codes used across the nation include a further subdivision of these

COAL T'WAS WEALTHS

Construction type
Occupancy or use
Apparatus and staffing
Life hazard

Terrain
Water supply
Auxiliary appliances
Street conditions

Weather
Exposures
Area and height
Location and extent of fire
Time
Hazardous materials
Special concerns

Figure 5-3 *An acronym can help a firefighter with scene size-up.*

classifications; however, for fire preplanning and strategy, the five NFPA classifications (Type I, II, III, IV, and V) will serve as a representative sample for the purposes of this text.

Type I or Fire-resistive Construction Type I construction is called fire-resistive construction, illustrated in Figure 5-4. The columns, beams, floors, walls, and roof are made of materials classified as noncombustible using the certification tests as specified in NFPA standard 251, *Standard Methods of Fire Tests of Building Construction and Materials.* This testing process is also recognized in the American Society for Testing and Materials (ASTM) E-119.

In fire-resistive construction, the structural members are of noncombustible materials that have a specified fire resistance. Type I is generally for buildings with more than one floor. The requirement is four-hour resistance of the columns and beams, three hours for the floor, and two hours for the roof materials.

Some important size-up considerations for Type I construction are the intensity of the fire, the durability of the fire-resistive materials used to protect the various structural components, the amount of hidden void spaces where fire can travel undetected, and the amount of open areas where fire can rapidly spread and undermine the resistance of the structure to collapse. These buildings are usually very well constructed and are structurally stable; if collapse occurs, it is generally localized to a specific area of the building.

Figure 5-4 *Type I or fire-resistive construction.*

Type II or Noncombustible Construction Type II construction is referred to as noncombustible construction (see Figure 5-5). It can be either protected or unprotected. In unprotected construction, the major components are noncombustible but do not have fire-resistant materials added for protection. The best example is unprotected steel (noncombustible) found as a roof support or steel beams that have not been protected. When exposed to fire, steel elongates, bends, and begins to lose its strength at approximately 1,000°F (537.8°C). In Type II construction, steel that is protected generally provides a degree of protection which is less resistive than Type I and is referred to as protected noncombustible construction.

Type III or Exterior Protected/Ordinary Construction The **ordinary construction** type of building has the exterior walls made of masonry materials (see Figure 5-6). The interior walls and materials are permitted to be partially or wholly combustible.

Ordinary construction
A type of building construction in which the exterior walls are usually made of masonry and therefore noncombustible.

(a)

Figure 5-5 *Type II or noncombustible construction.*

(b)

Figure 5-6 *Type III or ordinary construction.*

■ **Note**
Noncombustible buildings have many combustible features such as roofs, balconies, overhangs, and deck roofs. Nevertheless, firefighters need to understand that the designation of noncombustible does not indicate resistance to a collapse of the structure, because the contents of many of the buildings, especially plastic furnishings, add significantly to the fire load, which increases the amount of heat impacting structural components.

Fire load
The total amount of fuel that might be involved in a fire.

Heavy timber construction
A type of building construction where the exterior walls are usually made of masonry and therefore are noncombustible.

Even though the amount of combustibles per square foot is limited by the occupancy or building usage, the masonry exterior of these buildings can be deceptive, giving the impression of safety, while the interior can contain large amounts of combustible contents.

Type IV or Heavy Timber/Mill Construction Like the Type III construction, the exterior walls of a Type IV **heavy timber construction** building are masonry; however, the interior columns and beams are generally solid or laminated heavy wood and are designed and built without concealed spaces. Figure 5-7 illustrates an example of a Type IV, heavy timber/mill construction.

Because of initial design or through alteration, heavy timber buildings have many dangerous departures from ideal mill construction. Type IV construction

Figure 5-7 *Type IV or heavy timber/ mill construction.*

is often called "slow burning," which is advantageous when an interior attack can be conducted; but once a exterior defensive attack is underway, a better description might be a lengthy burn producing high amounts of BTUs.

Type V or Wood Frame Construction In Type V construction, all major structural components can be made of combustible materials as shown in Figure 5-8. The primary support for the building is from the wood components with little or no fire-resistive protection. A major fire problem is the concealed voids and channels formed inside the walls between the wooden wall studs as well as the opportunity for fire spread to the upper portions of the building.

Even though modern wood-frame buildings may be structurally considered platform frame buildings, they do contain void areas such as **soffits, utility chases,** and truss floors and ceilings which are interconnected by electrical and plumbing areas. These areas provide a path for fire and heat movement, and thus create a dangerous condition for firefighters.

Occupancy or Use

In size-up, the **occupancy or use** of the building can provide information about a building's use and contents. Loss of lives in building fires is always a concern where large numbers of people gather.

Apparatus and Staffing

Knowing the manpower resources responding on the initial alarm assignment—including the number, type of pumping, and ladder capability—is important. By knowing and understanding these limitations, additional needed resources can be requested using experience or departmental **standard operating procedures (SOPs)** as a basis to determine the type and amount needed.

Soffits
False spaces in the underside of a stairway and projecting roof eaves or the false space above built-in cabinets generally in kitchens or bathrooms.

Utility chase
A channel used for electrical, telephone, and plumbing lines and pipes for the various services in the building.

Occupancy or use
The building code that classifies buildings by their use.

Standard operating procedure (SOP)
Specific information and instructions on how a task or assignment is to be accomplished.

Figure 5-8 *In Type V construction, all major structural components can be made of combustible materials.*

■ **Note**

Firefighters need to be concerned with the structural integrity of these buildings as the structural strength of wood is diminished very quickly under fire conditions.

Personal alert safety system (PASS)
A small, motion-sensitive unit attached to and worn with the SCBA by firefighters when entering an Immediately Dangerous to Life and Health (IDLH) environment.

Life Hazard

The priorities of the fire department at the fire scene are protection of life, fire containment, and property protection. To save lives, it sometimes becomes necessary to conduct an aggressive primary search for fire victims during the first few moments after arrival. This process is referred to as the primary search.

The entry should be by a search team properly equipped with full protective clothing, including a breathing apparatus with a **personal alert safety system (PASS)** in place and in operation. The PASS device detects nonmovement of the wearer and if motion stops for longer than thirty seconds, the unit sounds a loud audible alarm to assist other firefighters in locating the motionless firefighter. Older PASS devices required firefighters to turn them on and some firefighters failed to do so. Newer devices are automatically turned on when the air bottle valve is activated.

The important aspect of the primary search is that it is conducted in a systematic manner. The pattern of search can vary but the search pattern needs to cover all the areas of the structure including areas behind doors, closets, and other locations where an individual may have attempted to find safety from the heat and hot gases. See Chapter 6 for a discussion of overhaul and salvage work, as part of these activities is looking for persons who may be missing.

Terrain

The terrain or the ground level at which a building is built is important to firefighters since the structure can be built on land with different grade levels (see Figure 5-9). The front side of the building may be only two stories from street

Figure 5-9 *Variation of building heights.*

■ **Note**

Although these buildings present a heavy fire load because of the amount of solid combustible materials used in the construction, the amount of combustible contents is restricted, which keeps the fire load in the building interior at a minimum. Fire in these buildings that is well underway should be considered as a defensive fire.

level or ground level, whereas the rear of the building could be as high as four or more stories from ground level.

In some situations, buildings are set back from the street making it difficult for emergency apparatus to access the structure. In other cases, buildings may have fences, trees, or electrical wires in front of the building. These things can make access difficult and time consuming. Careful preincident planning can help in reducing the time required to deal with these problems.

Water Supply

Water supply must be thought of in terms of whether it can be delivered in sufficient gallons per minute to suppress the number of BTUs being given off by the fire. The water supply evaluation should be part of preplanning. For example, preplanning may include learning the location of water mains, hydrants, their sizes, and the availability of water from lakes, ponds, or rivers. Another consideration for adequate water is the location of water lines of other jurisdictions or districts where interconnecting valves could be used to redirect additional water during an emergency.

Auxiliary Appliances

The presence or absence of auxiliary fire protection systems is important in the size-up process. See Figure 5-10 for an example of a wet pipe sprinkler system, and Figure 5-11 for an example of a dry pipe sprinkler system. Fire sprinklers have a record of over one hundred years of being 90% to 98% effective in controlling fires to a point where the fire is not allowed to extend beyond its area of origin (NFPA 13, 2002 ed.). The majority of sprinkler ineffectiveness incidents (the 2% to 10% where they failed to work) are related to the closing of sprinkler valves, explosions, or flash fire. Three types of water sprinkler application systems are available.

1. Wet pipe
2. Dry pipe
3. Deluge

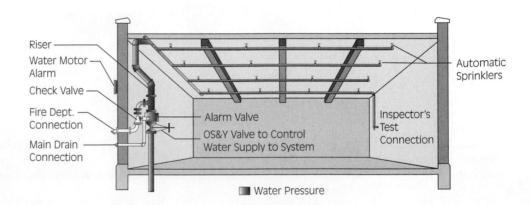

Figure 5-10 *A wet pipe has water in the system at all times.*

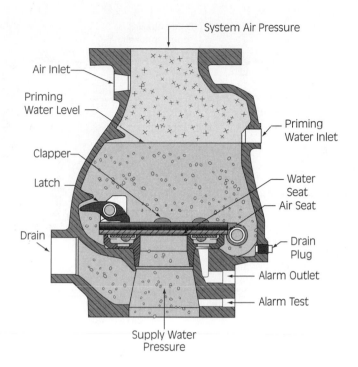

System Air Pressure

Air Inlet

Priming Water Level

Priming Water Inlet

Clapper

Latch

Water Seat

Air Seat

Drain

Drain Plug

Alarm Outlet

Alarm Test

Supply Water Pressure

Figure 5-11 *A dry pipe system contains water only when the system is activated.*

Deluge system

A system designed to protect areas that may have fast-spreading fire engulfing the entire area. All of its sprinkler heads are open, and the piping contains atmospheric air.

Standpipes

A manual fire fighting system with piping and hose connections inside buildings.

Dry pipe systems are generally installed in locations where broken pipes or malfunctioning sprinkler heads release water when no fire exits.

A **deluge system,** shown in Figure 5-12, is designed to protect areas that may have fast-spreading fire engulfing the entire area. All of its sprinkler heads are open, and the piping contains atmospheric air. Upon system operation, water flows to all heads providing complete water coverage. The system has a deluge valve that opens when activated by a separate fire detection system.

Some buildings are equipped with **standpipes,** most of which are designed for fire department use. Standpipes are designed to deliver water for manual fire fighting inside buildings. Firefighters can use these systems rather than pull hose lines into the building. Figure 5-13 illustrates the standpipe systems found inside of structures.

Street Conditions

Street conditions such as narrow streets, traffic congestion, double-parked cars, and construction work can severely impact fire operations. In some cases it may not be possible to get the fire apparatus close to the actual fire building. All of these factors impact the ability of the arriving units to attack the fire quickly. Close coordination with the public works and traffic departments can reduce such problems.

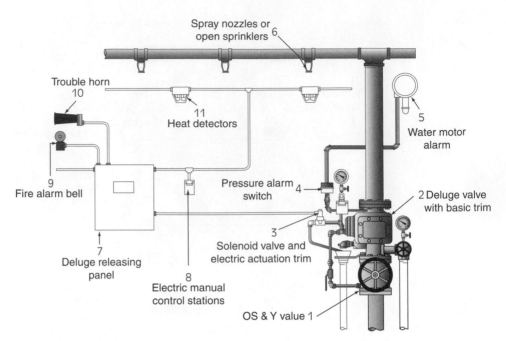

Figure 5-12 *Deluge system.*

Weather

Temperatures that drop below freezing can result in very slow emergency operations, as shown in Figure 5-14. Ice accumulation on hose lines, ladders, and protective clothing makes movement slow, exhausting, and dangerous. Additionally, severe cold causes mechanical failures such as hydrant freezing and sluggish or malfunctioning apparatus operation.

High temperatures and high or low relative humidity conditions also seriously affect the amount of work firefighters can do, because protective clothing holds in body heat and thus increases body temperature. This increase in body temperature also increases the body's fluid requirements, so firefighters must drink additional fluids. The increase in body temperature can reduce the amount of work potential and lead to heat exhaustion or heat stroke. (See Table 5-1 for details.)

The key to successful operations under extreme weather conditions is to ensure working crews have in place a **rehabilitation** system containing:

- hydration;
- nourishment;
- rest and recovery; and
- medical evaluation.

High temperatures also raise the temperature of fuels such as wood, flammable liquids, and other substances. This increase in fuel temperature brings the fuel closer to its ignition temperature, thus reducing the temperature needed for ignition of the material or substance.

Rehabilitation
A group of activities that ensures the health and safety of responders at emergency incidents.

(a)

(b)

Figure 5-13 *An example of a Class I standpipe system (a) and a Class II standpipe system (b).*

High humidity (the percentage of water in the air) can and does impact the burning process. An increase in moisture content in the air impacts the movement of the smoke and heated gases as the more humid air has a tendency to be heavier than less humid air and the smoke and heated fire gases will then tend

		TEMPERATURE °F											
		45	40	35	30	25	20	15	10	5	0	−5	−10	−15

WIND SPEED (MPH)		45	40	35	30	25	20	15	10	5	0	−5	−10	−15
	5	43	37	32	27	22	16	11	6	0	−5	−10	−15	−21
	10	34	28	22	16	10	3	−3	−9	−15	−22	−27	−34	−40
	15	29	23	16	9	2	−5	−11	−18	−25	−31	−38	−45	−51
	20	26	19	12	4	−3	−10	−17	−24	−31	−39	−46	−53	−60
	25	23	16	8	1	−7	−15	−22	−29	−36	−44	−51	−59	−66
	30	21	13	6	−2	−10	−18	−25	−33	−41	−49	−56	−64	−71
	35	20	12	4	−4	−12	−20	−27	−35	−43	−52	−58	−67	−75
	40	19	11	3	−5	−13	−21	−29	−37	−45	−53	−60	−69	−76
	45	18	10	2	−6	−14	−22	−30	−38	−46	−54	−62	−70	−78

A   B   C

	WIND CHILL TEMPERATURE °F	DANGER
A	ABOVE 25°F	LITTLE DANGER FOR PROPERLY CLOTHED PERSON
B	25°F / 75°F	INCREASING DANGER, FLESH MAY FREEZE
C	BELOW 75°F	GREAT DANGER, FLESH MAY FREEZE IN 30 SECONDS

Figure 5-14
Subfreezing temperatures can slow emergency operations.

to move horizontally. Vertical as well as horizontal ventilation may be needed to speed up the ventilation process.

Exposures

Exposure
A property endangered by radiant heat from a fire in another structure or an outside fire.

Water does not absorb radiant heat well, as most of the radiant heat will pass directly through the water spray. Using water for **exposure** protection against radiating heat transfer is best accomplished by running water down the side of the exposed building, coating it with water. This way, the water transfers the built-up heat in the building surface by conduction as it flows down the surface of the building.

In addition, water spray can protect against the movement of heated air currents (convection) and direct flame impingement. Firefighters use large water appliances to reduce or redirect air currents and to cool threatened building surfaces.

Area and Height

The area and height of the building are concerns as they indicate the maximum potential fire area. It is important for firefighters to check the entire perimeter of the building before committing resources inside the building as the total size and area can be deceiving when looking from outside only.

Table 5-1 *Heat stress index developed by the U.S. Fire Administration.*

	RELATIVE HUMIDITY								
TEMPERATURE °F	10%	20%	30%	40%	50%	60%	70%	80%	90%
104	98	104	110	120	132				
102	97	101	108	117	125				
100	95	99	105	110	120	132			
98	93	97	101	106	110	125			
96	91	95	98	104	108	120	128		
94	89	93	95	100	105	111	122		
92	87	90	92	96	100	106	115	122	
90	85	88	90	92	96	100	106	114	122
88	82	86	87	89	93	95	100	106	115
86	80	84	85	87	90	92	96	100	109
84	78	81	83	85	86	89	91	95	99
82	77	79	80	81	84	86	89	91	95
80	75	77	78	79	81	83	85	86	89
78	72	75	77	78	79	80	81	83	85
76	70	72	75	76	77	77	77	78	79
74	68	70	73	74	75	75	75	76	77

NOTE: Add 10° F when protective clothing is worn and add 108F when in direct sunlight.

HUMITURE °F	DANGER CATEGORY	INJURY THREAT
BELOW 60°	NONE	LITTLE OR NO DANGER UNDER NORMAL CIRCUMSTANCES
80°–90°	CAUTION	FATIGUE POSSIBLE IF EXPOSURE IS PROLONGED AND THERE IS PHYSICAL ACTIVITY
90°–105°	EXTREME CAUTION	HEAT CRAMPS AND HEAT EXHAUSTION POSSIBLE IF EXPOSURE IS PROLONGED AND THERE IS PHYSICAL ACTIVITY
105°–130°	DANGER	HEAT CRAMPS OR EXHAUSTION LIKELY, HEAT STROKE POSSIBLE IF EXPOSURE IS PROLONGED AND THERE IS PHYSICAL ACTIVITY
ABOVE 130°	EXTREME DANGER	HEAT STROKE IMMINENT!

■ Note

The height of a building is a factor in how far the walls will collapse. Establish a collapse zone and have a preplanned warning system in place.

Collapse zone
An area next to the structure generally 1.5 times the height of the building. Personnel are not allowed to operate within this zone.

Personnel accountability system (PAS)
A system set up to determine the entry and exit of personnel into the working area of an emergency incident.

Rapid intervention team or crew (RIT/RIC)
The assignment of a group of resources with the sole purpose of rapid deployment to reports of operating personnel in trouble or missing.

Personnel accountability report (PAR)
Periodic reports on the location and condition of personnel who have entered the working area of an emergency incident.

The height of the building raises issues of whether the fire department has the necessary ladder lengths to reach upper building areas and if the height provides an exposure hazard to nearby buildings. The height also impacts the use of master streams. If the building height is less than four floors, outside master streams can in some cases penetrate the lower building floors.

Location and Extent of Fire

Generally the lower the fire is located in the building the more serious the threat of fire is to the building. A low building fire exposes most of the building to the upward movement of the fire and heated fire gases.

A second concern is a fire below grade such as a basement fire, subway tunnel fire, or one below deck on a ship. These fires are hotter and generally more complex than the same fire above ground level, because there are few ways to ventilate them horizontally and the stairs in the lower basement area provide a vent channel for the hot fire gases to travel upward. This requires that the fire attack itself be advanced through this hot, smoky stairwell to get to the seat of the fire with little or no ventilation. This dangerous situation requires all safety precautions to be in place before the attack.

Time

The time of the incident provides firefighters with some insight into whether they may encounter a life-threatening fire. The life hazard in residential properties where people sleep is far greater at night than during the daytime. The time of day also impacts the time required for a fire apparatus to arrive as morning and evening traffic peaks can double the response time. In certain occupancies, the number of people will vary with the time of day or the day of the week. For example, an incident at a school at 10:00 A.M. on a weekday may be a serious life problem, whereas at 2:00 A.M. on Sunday or on a holiday, the life threat concern is not as great.

Special Concerns

There are other special concerns needing to be addressed while sizing up the emergency. The first concern is the need for a **personnel accountability system (PAS)**, which is a system to track firefighters entering and leaving the immediate fire attack zone so that all members can be located on the incident at any point in time. Second, a **rapid intervention team or crew (RIT/RIC)** must be in place to respond to incidents involving firefighters who become trapped or disoriented in the structure. The third area of concern is when the situation deteriorates rapidly. In this case, a **personnel accountability report (PAR)** becomes necessary. When called for, all personnel must immediately report their status. Some departments require a PAR on set time intervals during an incident.

FIRE BEHAVIOR IN SPECIFIC OCCUPANCIES

Fire problems and fire behavior patterns are identified by firefighters for selected occupancies. Different procedures need to be followed for various types of buildings. It is important to understand the essential differences of building methods and building types when evaluating how to approach a fire.

Building Construction Methods and Occupancy Types

Platform construction method

The floors in a building are built separately from the outer walls; thus, the ceiling and floor area serves as a fire block to stop the movement of hot fire gases between floors.

Balloon frame method of construction

An obsolete construction method in which the wood studs run from the foundation to the roof and the floors are nailed to the studs.

In a building constructed using the **platform construction method,** the floors in a building are built separately, thus in the outer walls the top of the ceiling and floor serve as a fire block to stop the movement of hot fire gases between floors (see Figure 5-15). In most cases, this method of construction will impede the upward movement of hot fire gases. However, utility chases (channels to carry utilities such as phone, electrical, and plumbing) and poke-through construction (where contractors or owners poke holes in existing fire-resistive walls to add or to rerun electrical, phone, or plumbing lines) provide channels for the movement of hot fire gases throughout the building.

The **balloon frame method of construction** is where the floor is attached to the outside of the wall studs. This configuration does not provide fire stoppage between the floors and, as a consequence, a fire started in a basement of the building can quickly spread up the vertical channels of the wall and reenter the building on the upper floors. Conversely, floors in buildings of platform construction are built separately and therefore serve as a fire block.

Figure 5-15 *Utility chase.*

Single-Story Family Dwellings of the Past

The types of fuels found in most residential structures today have dramatically changed from those during the 1940s and 1950s. A typical residential property of that period had plastered walls covering wood or wire mesh which was painted or covered with heavy paper or cloth. The floors were covered with bare wood, rugs, or wool carpets. Some were covered with linoleum. The furniture varied from bare wood to upholstered furniture with cotton, wool, or leather. Latex rubber was found only in some cushions and pillows, and the mattresses were filled with either cotton stuffing or feathers. The rooms were poorly insulated with single-paned windows which allowed air to freely circulate. The fire load was low and the rooms were well ventilated.

Today's Residential Properties

The fire load and fire behavior problems encountered in residential properties have significantly changed over the last two decades. First, these properties are better insulated, including the installation of double-paned glass to better insulate the interior from both heat and cold. This insulation holds in the heat from an interior fire. Second, residents tend to use more plastics or products made from hydrocarbons, such as plastic furniture, decorations, carpets, and window coverings.

The combination of the superior insulation and the increase in the total fire load has made interior fire fighting hotter, has decreased the time to flashover, and in general has become much more dangerous in these occupancies. Firefighters should expect a wide variety in the intensity of fires in these occupancies.

Fires encountered in one-story residences vary from those that can be extinguished with a garden hose to those that involve the entire building. The fire service responds to more single-room fires, as they are more common than a fully involved dwelling fire. Single-room fires can generally be extinguished with a 1.75-inch (45-mm) hose line, backed up with a larger line to protect against a fast-spreading fire where additional water may be needed. The line is brought into the interior of the dwelling from the uninvolved area with the objective of confining the fire area of origin.

A fully involved one-story dwelling of average size can generally be attacked with two 1.75-inch (45-mm) lines. The primary extinguishment objective with a totally involved structure is to keep exposures from becoming involved. In some areas of the United States, dwellings are built close together. If wood is used as siding, then the exposure problem can be severe.

On a large dwelling fire, it is a good practice to start the attack using 2.5-inch (64-mm) lines. Some larger, single-family house fires can be suppressed with only one of these larger lines. Once extinguishment is achieved, these larger lines can be reduced to 1.75-inch (45-mm) lines. These smaller lines are more maneuverable and they help to reduce water damage. Because the 1.75-inch (45-mm) lines are more maneuverable, their use allows personnel to be released for work in other areas.

Multiple-family Dwellings

Living arrangements in multiple-family buildings vary from city to city as does building construction, resulting in different fire problems. In some cities these dwellings are three-story buildings built with wood frame construction. A separate family resides on each floor.

In contrast, other cities have three apartment flats each containing three stories with one family per floor with rear porches on each floor. Most of them were built with brick wall construction with weak masonry work and open attic areas. These apartments using the same construction methods require firefighters to always check for signs of a possible building or wall collapse. Figure 5-16 depicts the common open attic with truss floors often found in these buildings.

Commercial Fires

Most fires in mercantile stores occur in a ground floor store or the basement located underneath the store itself. If the property is not protected with fire sprinklers, fire can travel rapidly because large amounts of combustible goods are stored there. Always check for the horizontal spread of fire as fire-resistant separations may not be present. Furthermore, if this material was installed at one time, it may now contain openings or holes allowing fire and smoke to spread.

The subdivision of the basement into areas or rooms increases the difficulty for firefighters as they will encounter a very hot, smoky fire that is rapidly spreading upward. Because of the extreme heat and smoke, advancing a line into the basement in most cases also is not possible. Fire attack must be conducted through an opening in the upper ceiling area of the basement. From the floor above the fire, a cellar nozzle is placed in the floor or the basement ceiling; this special nozzle is designed to disperse the water in a circular pattern at various room levels.

Figure 5-16 *Open attic and truss construction.*

Mercantile Fires

A common type of mercantile building that presents difficult fire fighting problems is the so-called strip mall. The name derives from the fact that the building is built with speculation in mind, meaning the owner has the building built with the idea that it will be quickly sold in an effort to make a large profit. As a result, contractors adhered to the minimum construction requirements, building the structure as cheaply as possible. As a result, fire protection of these buildings barely meets the building code requirements. Strip malls are usually built as a row of stores, one story high, but in some cases, two-story buildings are erected with stores on the ground floor and other occupancies above. Construction is usually of lightweight materials with poor workmanship. An open **cockloft** frequently covers the ceiling area of several stores.

A few strip malls have fire-resistant walls, but generally not between every store. When the walls are present, they are often inadequate. In some strip malls there is a common basement that extends under the row of stores. In most cases, storage space for each store is separated only by chicken wire or wood-slat partitions, allowing fire to spread horizontally.

Because of the type of construction, a fire originating in one store of a strip mall has a good chance of communicating to others as it can extend into the cockloft area and once there, will move horizontally with little or no resistance.

Ventilation at the roof is important. A large hole should be cut to draw and release the heated fire gases out and delay the horizontal travel. It is recommended that the hole be cut directly over the fire, and if possible, it should be about 80 ft^2. It is also desirable to get at least a small hole in the roof before the ceilings are opened, to prevent a backdraft condition. The object is to get in front of the fire to be successful.

Hotel Fires

Experience with hotel fires tells us that these buildings were not designed with fire safety in mind. Many were built with vertical openings inside the building to allow the vertical movement of fire into the upper portions where occupants sleep.

Older hotels were built with open stairways and rooms off of corridors using windows over the doorway to provide ventilation for the room. These windows, called **transoms,** allowed fire and hot gases traveling in the hallway to enter the room through the upper levels of the corridor. These older windows are being replaced with central air and heating units which require the room and hallway to be enclosed.

Central **heating, ventilation, and air conditioning (HVAC) systems** present a different problem. These units can circulate hot fire gases throughout the building if the unit is not protected with **fire dampers.** Fire dampers are used to prevent movement of heated fire gases in the ducting system. They use a fire-resistant shutter that is operated by a fusible link inside the duct to stop the movement of the gases. Some newer HVAC systems automatically shut down the fan in the event of fire to prevent the air and smoke in the duct from being circulated

Cockloft
The void space, approximately 3 ft deep, between the ceiling area and the underside of the roof.

Transom
A small window located in an older building, such as a hotel, at the top of the ceiling, generally located over the room entrance to allow heated air to circulate into the room as most were not equipped with heating units in the room.

Heating, ventilation, and air conditioning (HVAC) system
A central system used to heat and cool large buildings.

Fire dampers
Items used to prevent transmission of flame where air ducts penetrate fire barriers.

throughout the building. If the system is not automatic, it is important that firefighters shut down the system to prevent the spread of fire. Some HVAC systems are reversible, meaning that the fans can be set to remove smoke from inside the building. The operational instructions and location of the HVAC system(s) should be identified during fire preplanning sessions.

A very serious fire behavior problem encountered at hotels, apartments, and high-rise buildings is that they are not equipped with an **eyebrow,** a concrete extension over the top of openings such as the windows, doors, and balcony areas. Eyebrows are designed to prevent or inhibit fire and smoke from lapping into the upper floor(s) in a multistory building. This lapping of fire and heated fire gases on the exterior of the building is sometimes referred to as **auto-exposure.** The concrete extension prevents flames from lapping out of the window from one floor into the next higher floor, spreading the fire.

Eyebrow
A concrete extension over the openings on a multistory building to prevent the expansion of fire from lower rooms to the upper floors.

Auto-exposure
The lapping of fire from one floor to the upper floor on the exterior of the building.

Industrial Occupancies

Industrial buildings vary considerably because manufacturing processes differ widely resulting in a mix of hazards and concerns to firefighters. A number of hazardous materials and processes can be found in these buildings, which require preplanning to identify. Hazardous materials and operations can be found in both old and new industrial buildings.

One of the common types of older industrial buildings is the tenant factory or loft building. This non-fire-resistant building is generally built four to six stories in height with brick and wood joist construction. The building fronts vary from 20 to 60 ft (6.1 to 18.3 m) in width and from 40 to 200 ft (12.2 to 61 m) in depth. In many cases, the front is built between two streets. There is usually one factory per floor, but in some cases the buildings will have two occupants on each floor. Although there is a tendency for manufacturers of similar products such as clothing to congregate in one building, it is quite common to find several completely unrelated products being produced in the same building.

In these occupancies the potential for a fast-spreading fire is high because of large amounts of combustible goods and products, or perhaps manufacturing processes that involve the use of heat or flame.

A few of the older-type factory buildings have open stairs and unprotected light and ventilation shafts. In a number of them the elevator shafts are open with large skylights. The skylights are all aligned to allow light into the work area. If the skylight is broken or open, the fire will vent and result in a rapidly spreading fire that burns with great intensity. Large hose lines with copious amounts of water are needed to stop the fire from spreading and to protect exposures.

Newer manufacturing buildings are constructed using concrete tilt-up construction. Some are built with prepoured concrete panels, which are either raised at the job site or poured at the site. The walls are raised and then held together at the roof level with prefabricated wood trusses. They are then covered with a lightweight wood sheeting material.

The trusses provide open void spaces where fire gases can accumulate. The truss components are connected with lightweight metal connectors, which are not protected with a fire-resistant material. Fire quickly weakens the metal truss connections or wooden trusses which can lead to the collapse of the roof. In one recorded fire, the roof truss sections fell, killing five firefighters in an industrial building that was used as an automotive repair garage.

Churches

Although services may vary considerably among religious groups, churches and other buildings where services are conducted are remarkably similar in construction. Therefore, when we refer to church fires, we include any building used regularly for religious services fitting the occupancy description.

In church fires, the construction feature that usually leads to the church's destruction is a large hanging ceiling or cockloft. Once the fire reaches this area, it will burn without interference because the area is virtually inaccessible. Generally the ceiling is too high to be opened from below, the walls are too thick to be pierced, and the roof is an unsafe area to work.

If the church has a steeple, chances are the fire will probably advance there and eventually result in its collapse. The height of the steeple may require aerial streams. Therefore, when placing the apparatus, firefighters should consider the distance the steeple may fall. Always protect personnel and equipment by providing enough space for wall collapse.

Many churches are interconnected to a rectory or living quarters by an unprotected passageway which could allow fire to travel (in either direction) unless prevented by large protective hose streams.

The life safety record for church fires is good, as very few have resulted in multiple deaths. This could be due to the fact that they are not frequently occupied and most church fires occur when the congregation is not present. Although the fire load from contents may not be high, church fires can be very difficult and dangerous as the roofs are generally steep with wide-open attic areas, which encourage rapid fire spread.

Schools

Although most states require schools to conduct practice fire drills and maintain fire-resistive construction standards, we are still faced with fatal school fires. In recent years, there has been a significant improvement in the use of fire-resistant construction materials in new schools. Nevertheless, firefighters still need to be concerned as a number of older schools are still in use.

Ventilation at school fires should be performed with the objective of directing the smoke and heat away from corridors and stairways. Hose lines should be placed in locations to support the protection of these vital escape routes.

■ **Note**

Problems facing an incident commander in a school fire are the confusion and the number of possible victims who may be trapped in the building. It is difficult for the first arriving companies to work through and find out the extent of the fire and the rescue problems at hand. A systematic search plan is needed and must be put into effect immediately by the incident commander. A number of pre-incident plan inspections and drills can greatly improve the actions of students.

■ **Note**

In firefighting, the fundamental rule is "heat rises," so it is important to be concerned about the vertical extension of fire. This is especially true of basement fires because the heat rises to the underside of the first floor. The heat is exposed to the floor joists, flooring, wall partitions, stairways, and dumbwaiters and open elevator shafts. All of these components provide a passageway to the upper portions of the building.

Drip loop (meter head)
The location where the electrical supply lines enter the building.

Schools may use modular classrooms. They are limited in height and required to have at least two-hour fire-resistant construction with special fire-resistant requirements imposed on the construction materials used for the walls, floors, and ceilings. They are normally located in groups of two to four units each with two exits for each classroom. These modular structures are relatively inexpensive and have an outstanding fire record.

Basement Fires

Basements are a good location for the storage of combustible items as they are generally located away from the main floor area. As a result, a fire in a basement can become fully developed and present a very hot, smoky fire situation where visibility is limited (see Chapter 3 on fire stages). In many cases, the basement will be stuffed with old furniture, other combustibles, and discarded items, significantly increasing the fire load and making advancement with the hose lines difficult.

Basements are usually large and some are not sectioned off, leaving a wide open area of combustibles that can produce a large, hot, fast-moving fire. Another problem facing firefighters is the distance between the basement entrances. In many cases, there is a great distance, which requires a long hose pull to get the nozzle within striking distance of the fire.

In some cases, an apartment basement is the area where the utilities enter the building. In this situation, there is a possibility of a gas ignition when the fire melts the piping connections at the gas meter or an explosion from a leak. It is essential that the gas service be shut off as soon as possible.

When the electrical service enters the apartment complex in the basement area, it is imperative that the electrical service be disconnected for the protection of working firefighters. In some cases, the electrical service can be shut off by cutting the electrical lines at the **drip loop** or **meter head** (see Figure 5-17) located on the outside of the building.

Attic Fires

Because fires burn upward, attic fires respond quickly to proper vertical ventilation. Attic fires should be attacked as fast as possible with a 1.75-inch (45-mm) line equipped with a fog or spray nozzle. The lines should be taken inside the building and advanced to the attic through the scuttle hole or attic stairway. If a scuttle hole or attic stairway cannot be readily found, then the ceiling should be opened from below and a ladder extended into the opening.

As little water as possible should be used to control the fire. Salvage operations should be started on the floor below the fire as soon as possible. If very little water is used there will be ample time available to spread salvage covers before the water comes through the ceiling. Figure 5-18 illustrates the ventilation of an attic and roof area.

Figure 5-17 *Drip loop or meter head.*

Figure 5-18 *Ventilation of attic and roof area.*

Flat Roofs

In a flat-roofed building, the attic space is approximately 3 ft deep and provides an open space between the underside of the roof and the top side of the ceiling. This open area allows fire gases to accumulate, thus heating the entire underside of the roof area. Several forces work to spread the fire. First, the heat from the fire gases preheat the combustible construction materials lowering the amount of heat needed for ignition. This increases the speed of fire spread. Second, when these flammable fire gases accumulate and are ignited, they produce a very-fast-spreading fire.

One flammable gas present at most fires is carbon monoxide (CO), an odorless, colorless, tasteless, flammable gas that has a wide explosive range from 12.5% to 74% with an ignition temperature of 1,204°F (651°C). It is produced by incomplete combustion and is mainly responsible for much of the fire spread encountered by firefighters. When conducting topside ventilation on a flat roof, always work with the wind at your back or the side from where the wind is blowing, called the **windward** side. This will keep the smoke and hot gases downwind from you. The opposite side is called the **leeward** side or the side to where the wind is blowing.

Windward
The side from which the fire is blowing.

Leeward
The side to which the wind is blowing.

Peaked Roofs

The steep inclines of peaked roofs encourage the use of aerial ladders as the ladder will provide a stable, flat work surface. However, aerial platform access is not always possible and the ventilation will need to be done using the roof ladder. The roof ladder is kept from slipping by securing the hooks over the roof ridge. In this position, the roof ladder evenly distributes the weight of the firefighter over a larger area of the roof. This is important to remember as a sloped or peaked roof is designed to carry less weight than a flat roof. In some cases, because the roof is designed to carry less weight, the dimensions of the lumber are often smaller than for the lumber used for a flat roof.

A vent hole is cut as close to the ridge as possible without damaging the underlying rafters. The cut should be as directly over the fire as possible, on the leeward side of the roof. Use a pike pole and make sure the vent hole is continuous through the ceiling into the area being vented.

Carport/Garage

A carport is an freestanding structure that is open on all sides. It is designed and built with a flat roof only large enough to keep the elements off the stored vehicle. Such carports are open and as a result the vehicles are subject to vandalism and arson. Multiple-vehicle fires have been encountered where a flammable liquid was used to ignite a number of these unprotected vehicles at the same time. This situation provides the opportunity for flammable liquid involvement from exposed fuel tanks.

This type of fire can be attacked using one 2.5-inch (64-mm) line or two 1.75-inch (45-mm) lines supported by an adequate water supply. A fast, direct attack with the hose nozzle is in most cases sufficient to quickly control these types of fires. The carport itself is not much of a fire problem because it is constructed using a minimum amount of combustible materials.

Attached garages may be found next to or under one-story and two-story residences. A one-story dwelling will require that a line be taken inside the house to prevent extension of the fire into this area. The two-story residence will require a direct attack on the fire in the garage area and a line upstairs to check the upward fire extension.

SUMMARY

Firefighters must understand the basic principles of fire fighting and the characteristics of fire behavior. There are five steps used in the decision-making process to determine the needed strategy and tactics of fire fighting. The acronym COAL T'WAS WEALTHS identifies specific factors that impact the strategy and tactics decision. The priority of tactical actions are in most cases set around three efforts: to locate the fire, to confine the fire, and to extinguish the fire. These efforts are made while also determining if the fire attack will be a direct mode, indirect mode, or nonattack mode. Life safety concerns for occupants and firefighters as well as concerns for the environment determine the mode.

Of importance are both general and specific fire situations of various occupancies, as well as suggested fire strategies and tactics to aid in successful fire fighting. Effective, efficient, and safe fire fighting applies knowledge of the fire behavior characteristics discussed in earlier chapters of this text and an understanding of fire behavior in various situations.

KEY TERMS

Auto-exposure   The lapping of fire from one floor to the upper floor on the exterior of the building.

Balloon frame method of construction   An obsolete construction method in which the wood studs run from the foundation to the roof and the floors are nailed to the studs. The resulting wall space provides a channel for hot fire gases to spread vertically to the attic area. See also platform construction method.

Cockloft   The void space, approximately 3 ft deep, between the ceiling area and the underside of the roof. This void space allows heated fire gases to move and spread fire into other parts of the building.

Collapse zone   The building collapse zone is an area next to the structure generally 1.5 times the height of the building. Personnel are not allowed to operate within this zone as the building may collapse. If structural collapse is expected, unmanned master streams are used to replace lines operated by personnel.

Decision-making model   A five-step process used to solve problems: (1) identify the problem(s), (2) identify the alternatives available, (3) select the best alternative, (4) implement the chosen solution, and (5) evaluate the implemented solution for expected results

Defensive mode   Generally a fire strategy which is conducted on the exterior of the building to protect the adjacent buildings from fire spreading.

Deluge system   A system designed to protect areas that may have fast-spreading fire engulfing the entire area. All of its sprinkler heads are open, and the piping contains atmospheric air. Upon system operation, water flows to all heads providing total water coverage. The system has a deluge valve that opens when activated by a separate fire detection system.

Drip loop (meter head)   The location where the electrical supply lines enter the building. A circuit breaker or electrical shutoff is generally found below the meter head at the circuit breaker box.

Exposure   A property endangered by radiant heat from a fire in another structure or an outside fire. Generally, property within 40 ft is considered an exposure risk, but larger fires can endanger property much further away.

Eyebrow   A concrete extension over the openings on multistory buildings to prevent the expansion of fire from lower rooms to the upper floors.

Fire dampers   Items used to prevent transmission of flame where air ducts penetrate fire barriers.

Fire load   The total amount of fuel that might be involved in a fire, as measured by the amount of heat that would evolve from its combustion expressed in BTUs.

Heating, ventilation, and air conditioning (HVAC) system   A central system used to heat and cool large buildings. They are of concern as they can spread hot fire gases if they are not equipped with fire dampers for control of the movement of the heated gases and smoke.

Heavy timber construction   A type of building construction where the exterior walls are usually made of masonry and therefore are noncombustible. The interior structures are large, unprotected wood members. To meet this definition, the columns must be at least 8 inches square and beams must be at least 6 inches by 10 inches.

Leeward   The side to which the wind is blowing.

Nonattack mode (passive approach)   Under certain circumstances, a fire attack may be too dangerous and the incident command will choose to let the fire burn out without an attack. Exposures may be protected at a safe distance.

Occupancy or use   The building code that classifies buildings by their use. Using this classification, the type of construction, size, and other fire and safety requirements are set forth in the appropriate codes.

Offensive mode   Fire fighting operations that make a direct attack on a fire for purposes of control and extinguishment. For structural fire fighting situations, this usually means interior fire fighting.

Ordinary construction   A type of building construction in which the exterior walls are usually made of masonry and therefore noncombustible. The interior structural members may be combustible or noncombustible.

Personal alert safety system (PASS)   A small, motion-sensitive unit attached to and worn with the SCBA by firefighters when entering an immediately dangerous to life and health (IDLH) environment. As long as the firefighter moves, the alarm does not sound. If movement stops for thirty seconds, the device sends a "chirping" warning sound; if the wearer fails to move, the device will go into alarm mode and emit a loud noise to signal that the wearer may be in trouble.

Personnel accountability report (PAR)   Some departments are required to report, generally by radio, when the situation deteriorates or some departments routinely require periodic reports on the location and condition of personnel who have entered the working area of an emergency incident.

Personnel accountability system (PAS)   A system set up to determine the entry and exit of personnel into the working area of an emergency incident. A number of methods are used by the fire service.

Platform construction method   The floors in a building are built separately from the outer walls; thus, the ceiling and floor area serves as a fire block to stop the movement of hot fire gases between floors. See *balloon frame method of construction*.

Rapid intervention team or crew (RIT/RIC) The assignment of a group of resources with the sole purpose of rapid deployment to reports of operating personnel in trouble or missing. See NFPA 1710 and 1720, and FEMA technical report 123.

RECEO-VS An acronym used when developing the strategy and tactics on the fire ground. It stands for *rescue, exposures, confinement, extinguishment, overhaul, ventilation,* and *salvage.* The first five are listed in order of priority; the last two may be used at any point to support the first five.

Rehabilitation A group of activities that ensures the health and safety of responders at emergency incidents. The activities include rest, medical care, hydration, and nourishment on an as-needed basis.

Size-up A method used by firefighters to identify the problem(s) presented by the incident. Once the problems are identified, size-up procedures are used to determine the best overall strategy to be used to solve the problem. The tactics are those actions needed to complete the strategy or plan.

Soffits False spaces in the underside of a stairway and projecting roof eaves or the false space above built-in cabinets generally in kitchens or bathrooms.

Standard operating procedure (SOP) Specific information and instructions on how a task or assignment is to be accomplished.

Standpipes A manual fire fighting system with piping and hose connections inside buildings. They are classified into three categories: *Class I system* used for manual fire fighting with a 2.5-inch (64-mm) hose connection, *Class II system* used for manual fire fighting with a 1.5-inch (38.1-mm) hose connection but as a first-aid application only, and *Class III system* used for manual fire fighting with a combination of 2.5-inch (64-mm) and 1.5-inch connections.

Transitional mode The critical process of shifting from the offensive to the defensive mode or from the defensive to the offensive mode. If not conducted safely, firefighters can be endangered as supporting hose lines and protected positions are being changed.

Transom A small window located in an older building, such as a hotel, at the top of the ceiling, generally located over the room entrance to allow heated air to circulate into the room as most were not equipped with heating units in the room.

Utility chase A channel used for electrical, telephone, and plumbing lines and pipes for the various services in the building. They provide a channel for the movement of fire and heated fire gases.

Windward The side from which the fire is blowing.

REVIEW QUESTIONS

1. Describe the five-step decision-making process and apply it to the size-up process.
2. Explain why the fire service uses an acronym to assist in the development of the strategies and tactics on structure fires.
3. Explain how the fire load of a building impacts fire behavior and extinguishment activities.
4. What fire behavior problems might a firefighter encounter on a church fire?
5. Describe the actions that you would take to vent a peaked roof. Why would you try to vent at the peak of the roof?

REFERENCES

American Society for Testing and Materials (ASTM), Report E-119.

Angle, James S. (2005). *Occupational Safety and Health in the Emergency Services* (2d ed.) (Clifton Park, NY: Delmar Cengage Learning).

Angle, James et al. (2008). *Firefighting Strategies and Tactics* (2d ed.) (Clifton Park, NY: Delmar Cengage Learning).

Brannigan, Francis L. (2008). *Building Construction for the Fire Service* (4th ed.) (Quincy, MA: National Fire Protection Association).

Clark, William E. (1986). *Fire Fighting/Principles & Practices* (New York: Dunn & Bradstreet, Technical Publishing—Fire Engineering).

Diamentes, David. (2007). *Fire Prevention: Inspection & Code Enforcement* (3d ed.) (Clifton Park, NY: Delmar Cengage Learning).

National Fire Protection Association. (2002). *National Fire Protection Handbook* (Qunicy, MA: NFPA).

———. Standard 251, Standard Methods of Fire Tests of Building Construction and Materials.

———. Standard 1710, Standard for the Organization and Deployment of Fire Suppression Operations, Emergency Medical Operations and Special Operations to the Public by Career Departments.

———. Standard 1720, Standard on Volunteer Fire Service Deployment.

Quintiere, James G. (1998). *Principles of Fire Behavior* (Albany, NY: Delmar Cengage Learning).

Roberts, Bill (Fire Chief, City of Austin). (1993, March). "The Austin Fire Department Staffing Study."

Smoke, Clinton. (2005). *Company Officer* (2d ed.) (Clifton Park, NY: Delmar Cengage Learning).

U.S. Department of Labor, Occupational Safety and Health Administration. Safety and health topics, *http://www.osha.gov/SLTC/index.html.*

U.S. Fire Administration. (1991). *High-rise Office Building Fire, One Meridian Plaza, Philadelphia, Pennsylvania.* Technical Report No. 49.

Chapter

6

SPECIAL CONCERNS IN FIRE FIGHTING

Learning Objectives

Upon completion of this chapter, you should be able to:

■ Explain prefire and postfire planning processes and describe how these activities will ensure safe, efficient, and effective fire fighting activities.

■ Describe fire behavior in confined enclosures with and without ventilation activities. Explain the various methods of ventilation and how each method impacts fire behavior.

■ Explain the activities of salvage and overhaul and their role in fire extinguishment, and methods used to reduce further property loss.

■ Explain the procedures used to ensure that utilities do not threaten the safety of the building or its occupants.

INTRODUCTION

This chapter brings together areas of special concern in fire fighting activities. These areas are prefire and postfire planning, ventilation, salvage, and overhaul activities. Special consideration is given to these areas to emphasize the importance of the basic understanding of fire behavior and how it can improve the efficiency, effectiveness, and safety of fire fighting activities. Discussing basic fire behavior principles during a preemergency inspection review of a property allows for a better prediction of how a fire at this specific location may behave. Likewise, by reviewing the behavior of the fire during the postincident conferences, firefighters can compare fire behavior principles to actual fire behavior on the fire ground in an effort to increase knowledge and improve future fire fighting activities. This same process is applied to ventilation, salvage, and overhaul activities.

Ventilation receives special review in this chapter, as improper ventilation can result in the death of firefighters, and can spread fire into other areas of the structure. Fire that spreads into other areas increases the threat to occupants, firefighters, and the entire structure.

Salvage operations are described as the protection of property that was not damaged by the fire or could be damaged by the fire, smoke, water, and related extinguishment activities. Salvaging is an important factor in fire fighting as these operations are commonly used by community members and peers as a measure of the professionalism of the firefighters.

Fire service professionalism begins with an understanding of the fire combustion processes and the application of those procedures to reduce the fire and smoke losses as much as possible by knowing how smoke, fire, and water move within buildings. This basic information can lower fire losses and increase fire ground safety.

The process of **overhauling** also gets special treatment because firefighters have been injured while conducting overhauling procedures because they may forget basic fire principles. This lapse of memory or having a moment of inattention can result in a fire flare-up that can seriously injure firefighters. Occasionally, fires that were thought to be extinguished rekindle, so we review some of the time-tested methods used to ensure extinguishment has been completed and accomplished safely.

ADVANCE PREPARATION FOR FIRE FIGHTING

The firefighter has a responsibility to preplan the community areas in his or her jurisdiction or area of responsibility (see Figure 6-1). In most cases, this area will be the first-in or first-due jurisdictional area of the engine company. The survey of the area seeks out those locations and occupancies where there is an elevated

---

Ventilation

A systematic process to enhance the removal of smoke and fire by-products to allow the entry of cooler air and facilitate rescue and fire fighting operations.

Salvage

The protection of property which was not damaged by the fire or could be damaged by the fire, smoke, water, and related extinguishment activities.

Overhaul

A systematic process of searching the fire scene for possible hidden fires or sparks that may rekindle.

Figure 6-1
Firefighters entering a building for preincident planning.

life threat, property threat, or because of special circumstances, a threat to responding firefighters. These locations or properties are then identified as target hazards. Once pinpointed, they are subject to a preplan inspection by each on-duty shift (or in volunteer fire departments at various times) so all members have the opportunity to inspect and review the hazards and the property.

Target hazards are those areas where the fire department may encounter an unusual hazard to life or property or a situation that may overtax the department's resources requiring additional resources from other fire departments.

The Preincident Plan Inspection and Review

The process involved in **preincident planning** will help further identify major fire threats within the community, as fire departments generally do not have the time to examine all properties and structures in their response areas. This activity allows firefighters to devote time to the structures or areas posing the greatest threat by prescribing what may be needed to meet future emergency incidents. It also facilitates carefully thought out plans for future emergency incidents.

Preincident planning is a process of planning fire fighting tactics and strategies or other emergency activities that can be anticipated to occur at a particular location or situation. Another reason for preincident planning is that firefighters should not wait to think out possible solutions to fire problems until they arise, as they know that the pressure of the emergency makes it difficult to think of all possibilities. A preplanning session allows careful, quiet reflection on the problem being addressed in advance of an emergency situation.

Figure 6-2 *A target hazard in a community may include a facility or property containing a high hazard storage area.*

Preincident planning

A process of preparing for operations at the scene of a given hazard or occupancy.

Fire management zone (FMZ)

A zone within a jurisdictional engine company's area where similar hazards are grouped by approximately equal needed fire flow and hazard.

Furthermore, the preplanning session often brings together the first-alarm companies, allowing all personnel to become familiar with the property by seeing and discussing the problems they may encounter.

Availability of Water

Water supply is one feature of prefire planning that is extremely important. There are several important aspects of the water supply that require attention. First is the location of the fire hydrants, as well as their flow capability and the water main size. In some departments, each engine company jurisdiction is divided into zones. A **fire management zone (FMZ)** is a geographical area where the water requirements and structure threats are averaged so the amount and type of resources sent on the alarm assignment are matched with the water requirement or needed fire flow and/or special hazard need (threat to life) within each zone.

Supplementary water sources are important too. Firefighters are responsible for being familiar with and prepared to use any supplementary water source. A prefire discussion with the water supplier should determine if other systems could be made available under emergency circumstances. Relays and water shuttles can be preplanned using water department equipment or other resources available under emergency conditions.

In some communities, static water supplies such as lakes, ponds, rivers, or small streams can be used to supply needed water. The resources, supplies, and methods can be discussed and any issues or needs can be determined and resolved before the incident occurs. Figure 6-3 illustrates that firefighters should note the location of any dry hydrants that will assist with getting water from static water sources.

Figure 6-3 *During preplanning, firefighters should note the availability of any dry hydrants at static water sources such as rivers, lakes, and ponds.*

Built-in Fire Protection

During prefire planning, all built-in fire protection equipment should be identified. The operation and procedures needed to start and shut off the system should be learned and reviewed as well as the locations of its key components. It is especially valuable to know where **Siamese connections** and **control valves** can be found when sprinkler systems are present as the fire department will be required to connect to and operate these systems. Figure 6-4 illustrates firefighters examining a sprinkler outside control valves during a preincident planning operation.

Siamese connection
A hose fitting designed with a double hose line connection that allows two hose lines to be connected into one (generally larger) line.

Control valves
Valves used to control the flow of water in sprinkler or standpipe systems.

Mutual Aid Resources

The use of neighboring community resources must be preapproved and carefully planned in advance. Firefighters need to know what types of resources are available and what the availability is of additional staffing as well as their capability and the procedures used by assisting fire agencies, so any issues of incompatibility can be worked out prior to a major incident.

In California, some fire departments use different hose threads than other areas of the state. This has led to some difficulties when attempting to use other jurisdictional fire equipment and hose with hydrant threads that are incompatible with national standard threads. This is just a small example of a problem that needs attention prior to a major incident.

Figure 6-4
Firefighters examine outside control valves during preincident planning.

Preincident Planning and Fire Behavior

When conducting preincident planning activities, it is helpful for firefighters to remember the principles of fire behavior as they review structures and their characteristics. The contents of certain rooms or areas may present unique challenges for firefighters if they are made of materials that will burn hotter or create toxic fumes. The layout of a building and the location of stairwells, roof hatches, skylights, windows, and a variety of building features can impact how firefighters can ventilate a structure in the event of fire. A working knowledge of fire behavior can be helpful to take along to preincident planning activities for considering tactical strategies and special hazards.

POSTFIRE ACTIVITIES

Postfire conference/ Postincident size-up
Following the incident, this analysis is conducted to exchange what was observed, the result of actions taken, and what lessons were learned so improvements can be made on future incidents.

It is imperative that the postfire planning session be conducted in a positive manner. A punitive session will cut off communication and the fire service leaders will lose valuable training opportunities. The negative aspects of the session may flow over into future activities on emergencies. An open, positive session will promote a learning environment where all members can build on the positive aspects of their actions as well as those areas needing improvement.

Postfire Conference

The **postfire conference** is conducted to improve future fire operations by using the lessons learned from the incident. The responding companies are generally invited to the conference, if scheduling allows. If not possible, several conferences may be necessary to cover the various work shifts or the conferences may need to be conducted at several different locations. It is very important that all members have an opportunity to participate in the discussions.

This participation generally is promoted by having each member describe his or her actions during the incident. This allows officers to gain a better understanding of all activities that occurred. Furthermore, it is important that the officer assigned to documentation of incidents attends all the conferences to document and gather the information being presented so it can be summarized and presented to all members. A summary describing the actions, both positive and those needing improvement, along with any proposed recommendations is highly recommended.

One way to direct the focus of the conference is to use a checklist of items that have been important on the fires encountered in the past. See the following textbox for a checklist containing questions to be answered during a conference.

Postincident Form

Origin

Were the possibilities of arson thoroughly investigated? Does the cause of the fire emphasize the need for increased fire prevention activities? If so, should the fire prevention inspections be conducted by station members or the fire prevention division or bureau?

Detection

Would the property benefit from the installation of automatic detection devices? If so, is the necessary legislation to require an automatic detection system possible? Are the costs of the automatic detection device cost effective?

Alarm

Was the alarm reported, received, and transmitted correctly? If not, what was the problem and can it be resolved so it will not happen again? Did the dispatcher accurately provide a clear, concise description of the fire/emergency when forwarding the information provided by the informant?

(continued)

Equipment and Staffing Response

Were the proper equipment and human resources dispatched? Were sufficient resources available? Were correct routes to the scene chosen?

Extent of the Fire

What was the extent of the fire upon arrival of the fire department? Did the fire continue to spread after fire fighting operations commenced? Did special circumstances delay a fast application of water?

Size-up

Did the commander use the acronym COAL T'WAS WEALTHS or another department-approved method to bring about consideration of all factors during the officer's size-up? What was the first arriving officer's report? Did it accurately depict the situation upon arrival? If not, how could it have been improved?

Preplanning

Did past preplanning activities and documents help the responders develop the best and safest strategy? Does the preplan for the structure need to be updated?

ICS

Was an incident command system put into operation? Was it correctly applied and used? Was the incident used as a training session for future fire officers?

RIT

Was a Rapid Intervention Team (RIT) put into place? Were they equipped and ready to respond? Were higher levels of teams needed? If not were they available?

Forcible Entry

Were there signs of a break-in prior to fire department arrival? Was the building opened in a manner so that damage was held to a minimum? Were special tools needed that were not available?

Rescue

What were the conditions upon arrival that impacted the rescue efforts?
Was adequate staffing available on the first alarm for successful rescue operations and was an accountability system put in place? Were rescue efforts

(continued)

successful? If so, what were the factors contributing to the success? If not, describe factors that contributed to the failure of the rescue efforts. Can these be resolved in order for future rescue efforts to be successful?

Ventilation

Was the building ventilated in a timely and correct manner? If not, what recommendations can be made? Was it a problem of size of ventilation hole(s) or location or some other issue?

Hose Streams

Were hose streams properly directed and applied? Were hose streams the proper size and pressure? Were any suggestions offered to improve application or ventilation issues? Were there problems advancing the hose line?

Salvage

Were salvage operations effective? Can an estimate of property saved or protected be made? Are there any suggestions for future steps that can be taken to improve salvage operations?

Overhaul

Was the fire completely extinguished? Was the building cleared of debris and excess water? Was sprinkler protection restored? Was the building properly closed up?

If arson was suspected, did the fire department stand by until the arson unit arrived?

Traffic Control

Did the police department assist in traffic control? Was the street opened for traffic as soon as possible? Were bystanders controlled?

Utilities

Were gas, water, and electrical services properly handled by the fire department? Was the utility service prompt?

Incident Communications

Were the communications on the incident clear and concise? Were the radio frequency assignments specific and logical and assigned per the incident communications form?

The checklist can be prepared as a departmental form and can be tailored to meet the needs of every department by adding or deleting items to better serve the community needs. Some departments require all members to fill in the needed information which then allows it to serve as a training aid.

Reviewing fire operations in a conference affords an opportunity for the members of the fire department who were not present at the fire to benefit from the experiences of those who were present. Conferences are also instructive for those members who were at the fire but whose activities were confined to a limited area or sector. It permits them to gain knowledge of the overall strategy, operations, and behavior of the fire.

Finally, a postfire conference is an opportunity to discuss how the operation could have been accomplished more effectively and how future operations can be improved. The following materials are suggested for a postfire conference.

- Accurate and complete information as it can be obtained
- A complete and accurate drawing of the fire as a visual aid, with photos to accompany the drawing, if possible
- A plan for a systematic discussion and explanation by personnel to develop the order events and operations on the fire ground

A postfire conference requires a great deal of tact and diplomacy to obtain maximum participation by those in attendance. Comments of a critical nature regarding an operation should be approached as a training process and not an organized way to place blame. Figure 6-5 shows an on-scene postincident conference being conducted. Information gathered is fresh in everyone's mind, allowing the analysis to be as accurate as possible.

Figure 6-5 *A quick on-scene postincident conference.*

Other postincident conference meetings may be conducted at the training center or other departmental facility at a later time to allow more participation and further discussion.

VENTILATION

Ventilation is extremely important to successful and safe fire fighting. Ventilation needs to be a high priority during fire size-up. Once the decision is made as to the location and method of ventilation, the actions to ventilate are immediately implemented. Although fire fighting techniques recommend ventilation activities should not be started until attack lines are in place, once the location for ventilation is located an opening should be started quickly. Generally, ventilation operations require more time than the placement and deployment of hose lines. Furthermore, where truck companies are used for ventilation, additional engine companies generally arrive and are operating prior to the arrival of the truck company.

In jurisdictions without truck companies these important assignments must be carefully worked out during the prefire planning sessions. During this process, an engine company is often predesignated for building ventilation. Companies assigned the ventilation function should be very familiar with fire behavior processes and the impact ventilation can have on the combustion process.

Roof Ventilation

Roof ventilation is very challenging as it can vary in difficulty. Ventilation operations can be as easy as removing a scuttle hole cover or as complex as manually opening a large area. One of the first considerations when preparing to ventilate a roof is to look for natural openings that can be used; however, these openings are of little value unless they are in the right location. If they are not in the correct location and are opened, they will pull fire to the opening. Some of the natural openings available for ventilation are scuttle holes, skylights, and stairwells.

Scuttle hole
An opening that allows entry into the attic area.

Scuttle Holes   A **scuttle hole** is an opening in the roof that leads into the attic. The hole is fitted with a cover that is sometimes locked. Scuttle holes can normally be used to ventilate only the attic, as they generally do not penetrate the ceiling area. When lifting off scuttle covers, it is good practice to have the wind to your back so that the heat and gases will be released in the direction away from you. A wooden cover can generally be removed by prying it open with an axe. Some covers have spring loaded metal covers that may be locked from the inside. These present a problem and in many cases it is easier to find another location to ventilate.

Skylights Skylights are used to provide sunlight to areas below the roof. Their value in the process of ventilation is the same as other building openings; it depends on what area they will ventilate. In hotels and apartment houses they can be used to ventilate the hallway of the top floor. In commercial buildings, they will generally ventilate the storage area. Skylights can generally be removed intact since the frames are normally only lightly secured at the corners. However, if any difficulty is encountered in removing them, and time is important, there should be no hesitation in breaking the glass. More modern versions of skylights are equipped with Plexiglas rather than wire glass. They provide automatic ventilation as the Plexiglas melts at approximately 400°F (204°C) and will create a passageway for ventilation to occur.

Stairwells and Elevator Shafts The use of a stairway that terminates on the roof for ventilation can be particularly valuable; however, it can also prove to be dangerous if improperly used. Using a stairwell for ventilation is relatively easy, as it only requires opening a door. Nevertheless, prior knowledge of the building can prove to be invaluable, as it is important to know what parts of the building below the roof will be threatened by opening the door.

Some elevators have an equipment room located directly over the elevator shaft. In some cases, these rooms have a door that can be quickly opened or removed for the release of smoke through the shaft to the outside environment. These equipment rooms can be located and discussed in the prefire planning process.

Sounding a roof
Using an axe or pike pole, firefighters search the roof ahead of them to determine if it is solid enough to handle the additional weight of firefighters.

Sounding the Roof The ventilation officer has the responsibility to conduct an evaluation of the roof to ensure that it is safe before firefighters are allowed on it. As firefighters enter the roof they should **sound the roof** ahead of their movement to locate any weak areas. An axe, **pike pole,** or **rubbish hook** can be used for the sounding. A good knowledge of roof construction is important in making this evaluation. It is highly recommended that buildings with truss roofs are fire preplanned and the fire safety concerns of truss roof construction be discussed with all members of the fire fighting team.

Some roofs by their nature are hazardous and require extreme caution when working on them. An example is a roof covered with 0.5-inch plywood in a commercial building which gives the appearance of being sturdy. Most roof coverings installed on buildings constructed prior to 1960 are solidly constructed and well supported. They are generally safe although they might give the impression of being somewhat springy. Extra care should be taken when working on buildings constructed after 1960 that are classified as lightweight construction.

Pike pole (rubbish hook)
A device with a sharp pointed head and a curved hook which is used to thrust into the ceiling area and then using the hook pull the ceiling materials downward.

Performing Roof Ventilation Factors that are of special importance when it is necessary to cut a hole in the roof. The ventilation hole needs to be cut directly over the fire. Sometimes, this location is readily identified by simple observation, by looking for discoloration of the roof, or by feeling the roof for the hot spot.

Cutting a hole in the ideal location means that the place where the hole is to be cut is directly over the hottest spot which is also the most dangerous place on the roof. If there is any doubt regarding the safety of personnel, then this spot should be avoided and a safer location should be found to make the cut.

On most large building fires, the hole can be cut within 20 ft of the hottest spot without seriously redirecting the hot gases and fire through uninvolved portions of the building. Pulling the fire through uninvolved portions of the building should be avoided; however, if there are lives threatened, firefighters will typically not hesitate to pull the fire away from the trapped persons.

Working on a roof directly over the fire is a dangerous operation. Many firefighters have fallen through the roof and have been killed during ventilation operations. When conditions are right, an elevated platform can be used as a safety factor. A line can be lowered from the platform and tied to the person opening the roof. The firefighter can be pulled to safety in the event of a roof collapse.

Louvering

Louvering
A method of roof ventilation that reduces the exposure to smoke and initially prevents cutting the supporting roof joists.

Louvering is an effective method for ventilating a roof, as shown in Figure 6-6. Properly performed, it helps reduce the exposure of personnel to smoke and heat as the roof is vented. It can be used effectively on sheathing but is most effective on plywood paneled decking.

The operation consists of making two longitudinal cuts on either side of a single rafter with the cuts parallel to one another. The longitudinal cuts are then intersected with cross cuts. Once the cuts are made, one side of the cut panel can be pushed down, resulting in a louvering effect. This method of opening the roof will ventilate the area below and direct most of the heat and smoke away from the working area. Cuts should be made so that firefighters making the cuts have the wind at their back and the louvered panel should be used to direct the smoke away from those making the cut.

Figure 6-6 *A finished louver opening for ventilation.*

HORIZONTAL OR CROSS VENTILATION

Cross ventilation is a means of using windows, doors, and other horizontal openings for ventilation. This method can be extremely effective whenever there are a large number of openings of sufficient size. It has the advantage of normally limiting the smoke to the floor being ventilated, which is particularly beneficial when life hazard is present or valuable merchandise in other areas of the building is threatened with smoke damage.

An evaluation of the wind direction should be made prior to commencing horizontal or cross-ventilation operations whenever an entire floor or a large area is to be ventilated. The operation will be most successful when the smoke and gases are channeled out the leeward side of the building. Once the direction of the wind is determined, double hung windows on the leeward side should be opened from the top and sliding aluminum or sliding wooden windows should be fully opened. Remove all drapes, shades, screens, and other obstructions that would impede the flow. Double hung windows on the windward side should be opened from both the top and bottom or from the top only after all the leeward windows have been opened. Sliding aluminum or sliding wooden windows should be fully opened. Again, all obstructions to airflow should be removed.

It is a good practice when cross ventilating to determine where the heated gases and fire will go once they leave openings. It is possible that they could extend upward and through open windows on the floors above or through openings into adjoining buildings. If there is such a possibility, then all exposed unprotected openings should be closed or spray streams used to cover these openings to prevent any extension of fire.

Sometimes it is desirable to cross ventilate when it is impossible to get the windows opened. Under these conditions, it may be necessary to break the windows. Care should be taken when breaking windows as both civilian onlookers and firefighters can be injured by flying glass.

Smoke fans can be useful in cross ventilation. They can be set up to create either a negative or positive airflow. The two types of smoke fans are **smoke ejectors** and **positive pressure ventilation** (see Figure 6-7).

Negative Pressure Ventilation

Negative pressure ventilation is a method of using mechanical fans to exhaust or pull the heated smoke and gases from the interior of the fire building and move it outside the building. Negative ventilation has been used for a number of years; however, it has proven to be a rather ineffective and inefficient method of smoke removal. The fans never seem to correctly fit the opening, and as a result provisions need to be made to block the opening around the sides of the fan unit. If a good fit is not found, the fan will cause **churning of the air,** which is the pulling

---

Smoke ejectors
Mechanical smoke fan designed to draw heated air and smoke outside of a structure.

Positive pressure ventilation
A process that uses mechanical fans to blow air into a structure to remove smoke and gases.

Negative pressure ventilation
A method of forced ventilation that pulls air and smoke out of a structure.

Churning of the air
The phenomenon of smoke being blown out at the top of the building opening, only to be drawn back into the structure at the bottom of the opening by the slight negative pressure (vacuum) created by the action of the mechanical fan.

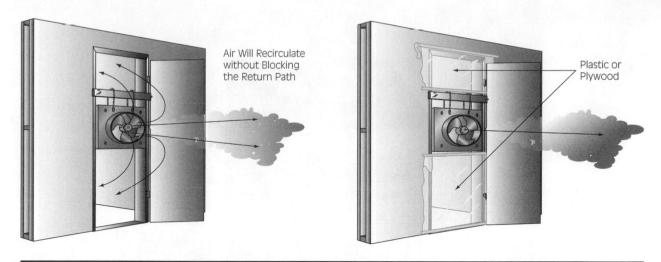

Figure 6-7   *While using smoke ejector fans for ventilation, firefighters must cover the entire opening to avoid churning of air which reduces smoke removal effectiveness.*

> ### ■ Note
>
> Care must be taken to minimize the movement of fire and smoke to building areas not directly impacted by the original fire.
>
> - Do not attempt to use the positive pressure ventilation for entry into the interior if the fire has progressed into the later stages or phases of the combustion process, a backdraft may result.
> - Do not push the smoke and fire into escape routes or into unburned areas of the building.
> - Keep the openings for smoke exhaust as close to the fire area as possible.

of air from the outside and drawing it into the building to be blown out again by the fan.

Other problems encountered on fires while using this method of ventilation are the picking up of debris, curtains, and other materials by the fan. They become trapped against the fan screen and need to be removed.

These fans were originally designed to be explosion proof, which enables them to be placed in an environment containing explosive gases; but in many cases, extensive use has damaged the electrical connections and the housing unit, making them dangerous in an explosive environment. For these reasons, the fire service tends to use positive ventilation.

Positive Pressure Ventilation

Positive pressure ventilation uses mechanical fans to blow air into a structure to remove smoke and gases through additional openings, allowing firefighters to gain entry for fire extinguishment. It combines the use of openings and mechanical blowers. To conduct successful positive pressure ventilation, the outlet opening must be controlled. If too many doors and windows are opened, positive pressure ventilation is ineffective, as the pressure in the smoke area will not be sufficient to move the smoke quickly and in the direction desired. For maximum effectiveness, good timing, the correct amount of opening space, and sufficient pressure is needed. The action of positive pressure ventilation is seen in Figure 6-8.

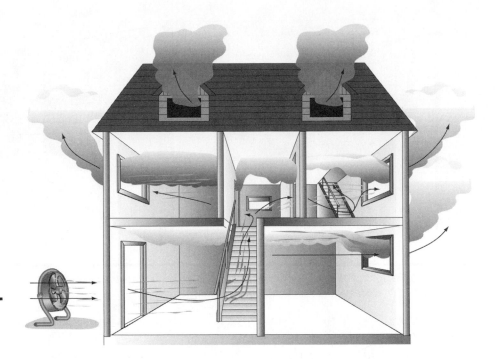

Figure 6-8 *Positive pressure ventilation.*

SALVAGE AND OVERHAUL

Historically, in the United States, fire fighting actions were based on bringing very large hose lines in the front of the building and pushing the fire from the front of the building back out the rear of the structure. Very little consideration was given to the amount of water and smoke damage or whether the fire department extended the fire damage in the structure. This resulted in the larger fire insurance companies funding their own salvage companies to respond to fires and work to keep the losses to a minimum. By providing this service, they could offset their losses and still pay for the costs of the salvage companies. However, as their operating costs increased, they phased out most of the salvage companies and added deficiency points to the insurance grading schedule for those fire departments not providing salvage work. Figure 6-9 shows one of the few remaining fire underwriters salvage patrols.

For a period of time, many fire departments did not recognize salvage and overhaul work as their responsibility. These situations resulted in delays in salvage and overhaul training and the purchase of salvage and overhaul equipment. The fire department now accepts the responsibility to extinguish fires and to reduce water and smoke loss as much as possible.

Figure 6-9 *New York Board of Fire Underwriters fire patrol.*

Overhaul Operations

The first step in the overhaul process is examining the behavior of the smoke, heat, and water from fire fighting operations and how it affected the building and its contents. Some materials readily absorb water, significantly increasing the **live load** in the building and threatening structural integrity. Smoke can damage or destroy clothing, draperies, and furniture while water can destroy drywall materials and some flooring products.

Structural integrity of the exterior walls and chimneys is important to the safety of the firefighters working inside and near the building. If needed, the walls or chimneys must be braced or shored up prior to any overhauling work. In some cases, removing excess water and water-soaked materials must be done as quickly as possible. The means by which water can be removed from buildings include using the building plumbing, diverting the water to outside the building by cutting or removing walls or floors, or building water chutes or containment areas.

Live load

Items inside a structure that are not attached or of a permanent nature.

To develop the best strategy, it is helpful to survey the damaged area before beginning overhaul operations. Part of the work survey should include a review of any tools or special equipment that might be needed in order to request them early and avoid a delay in getting the resources.

Once the survey is completed, the incident commander should attempt to contact the owner of the premises, if he or she is not at the site. This is particularly important if the fire is in a commercial occupancy, as it is always best if the owner is on the premises before overhaul operations commence. The owner can answer any questions regarding the inventory, business records, and the best way to safeguard the property.

The next step is the assignment of work tasks to be accomplished in teams. The team assignments are designated in a manner to keep the work of each group from interfering with the work of the other groups and to ensure all the work is completed in a timely manner.

Care must be taken to save all records that have been burned or partially damaged, as they may contain enough information to reconstruct the records needed to determine the inventory, the amount of insurance carried, and the financial condition of the business. The information contained in the records may be vital to the owner in establishing his or her loss and may be important information for arson investigators.

Debris Handling

Caution should be taken in removing burned items that were involved in a deep-seated fire from the upper floors of a building. If an elevator is used for this purpose, special procedures should be put in place. The object should be wrapped with materials that will prevent air from getting to the fire area. A water-type extinguisher is taken in the elevator. Firefighters have been surprised by a rekindle of a deep-seated fire as additional oxygen was supplied to the fire as the elevator descends. A handheld, water-based extinguisher can extinguish the fire sufficiently to allow firefighters to remove the burning material from the building.

Inside large industrial or mercantile occupancies, a safe location needs to be designated to conduct the final extinguishment. Special care in the selection of this location needs to be taken as materials moved into this area may be smothering. Additional oxygen can accelerate the fire.

All burned material needs to be thoroughly wetted down and sifted through thoroughly to make sure that no hot spots are present. In residential occupancies, it is generally better to move the debris outside of the building. The area where the debris will be piled should be chosen prior to any movement of the materials. It is a waste of time and effort to handle the burned material several times. If it is necessary to pile the materials on the sidewalk or street, adequate barricades must be provided and the responsible public agency should be notified.

Water Removal

One objective of the overhaul operation is to remove water from a building to prevent further damage, as it is possible for a building to become so overloaded with water that it could collapse the floor and walls.

Building collapse may occur at large fires where a number of heavy streams are projected into the building or at fires in sprinkler-equipped buildings where the sprinklers have been activated for a period of time. Occasionally, broken sprinkler risers or broken water pipes in the building will result in flooding. If this happens, it adds additional loss from water damage to the building and its contents, plus it increases the possibility of a building collapse.

Stairways can be used in multiple-story buildings for removing water. The removal is simplified if the stairs are concrete without a covering. If they are wooden or carpeted they can be protected with salvage covers. Salvage covers should be opened up to half their width to ensure a tight water seal.

Cast iron sewer pipes run both vertically and horizontally through buildings. This piping can be used for water removal. In many cases, the sewer piping system has a cleanout that can be opened to allow entry into the sewer system. If the cleanout is inaccessible, the piping can be broken open to enter the sewer system. It is important to provide a screening device so that the system does not become clogged with debris.

If there is no other means of removing water from a floor, it may be necessary to cut a hole in the floor. The hole should be cut close to an outside window and be of sufficient size to handle a large amount of water. All stock and other material located below the spot where the hole will be cut should be removed prior to performing this operation. A drain should be set up at this point which will divert the water to the exterior of the building. A ladder and salvage cover can be used for this purpose (see Figure 6-10).

A wall breach may be required in those cases where a large amount of water has collected on a floor and there are no other ways of removing it. This is performed by selecting an outside window with a windowsill close to the floor area where the direction of the flow of water on the exterior will not do further damage or reenter the building through the floors below. The window frame and glass should be removed and then the wall below it. The wall should be opened to allow the water to drain to the exterior of the building.

Checking for Lingering Fire

Thermal imaging device
A device that distinguishes between objects or areas with different temperatures and displays hot areas and cold areas as different colors on a video screen.

The infrared **thermal imaging device** sees a difference in temperature between objects or areas with different temperatures and provides a visual display of hot and cold areas. The differences can be seen on a video screen display device. Because the device is sensitive, the thermal imager can see a hot spot, even if the heat source is behind a wall. It is often used during search operations for fire victims, but also is an effective tool during overhaul operations.

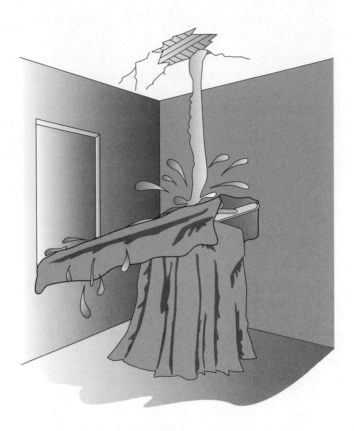

Figure 6-10 *A ladder with a salvage cover or plastic sheet can serve as a chute to divert water to the exterior of a building.*

Attic scuttle holes should be located and a firefighter sent into the attic to check for hot spots with instructions to include checking the area close to the eaves. Birds, rats, and squirrels from time to time build nests and store materials in these areas. These nests may have become ignited without direct flame contact.

If fire is suspected between floors, it is generally best to go below the suspicious spot and pull the ceiling down. Ceilings are easier to remove and less costly to repair than floors. If fire is suspected under the floor and it is impossible to get below the floor, then do not hesitate to cut a hole in the floor to gain access to the fire.

Fire has a tendency of getting under window facings, around doors, and under trim at floor level. If fire is suspected in these locations, baseboards and window facings around the doors and windows should be removed. When upper floors have been involved with fire, the bottoms of all vertical openings, including elevators, should be checked for burning debris.

Securing a Building

If possible the building should be turned over to the owner when the department is ready to leave. If the owner is not present or cannot be contacted, it then becomes necessary to secure the building against intruders. All windows and

Figure 6-11 *Dwelling boarded up after a fire incident.*

doors broken during firefighting operations should be boarded up. Lumber can usually be found on the premises for this purpose. If not, interior doors can be removed and nailed over openings.

The building should be made safe from weather elements before returning to the fire station. If holes were cut in the roof or if the fire burned through the roof, the holes should be covered with plastic covering materials. Figure 6-11 shows a structure that has been boarded up after a fire.

At some fires it may be necessary to post a firefighter as a fire watch. This assignment requires leaving a firefighter on the scene with a portable radio in order to quickly notify dispatch in case of a problem. If this is impractical, then the jurisdictional engine company or truck company should be assigned to check on conditions periodically. The assigned company must always stop and reenter the building to check for lingering fire.

SUMMARY

Understanding the basic principles of fire behavior and applying it to all fire fighting activities is extremely important for efficient, effective, and safe fire ground operations. Here we review the importance of these principles in prefire and postfire inspections, conferences, the procedures of ventilation, salvage, and overhaul to emphasize the importance of a continual update and application of the basic fire principles.

Prefire and postfire review and conferences provide the opportunity to review fire behavior principles before and after fires. They can also serve to reduce injuries and death as countless fire-fighters have been needlessly injured or killed due to unfamiliarity with the features of a structure or a hazardous situation. Likewise, firefighters have suffered from injuries and death from incidents where information gained from a postincident conference might have prevented their injury or death.

A variety of ventilation methods and the importance of proper ventilation procedures and how they impact fire behavior are discussed. If fire behavior is not properly controlled, firefighters and occupants can be killed or seriously injured.

Good salvage and overhaul procedures are also important. Fire and smoke damage losses and after-incident injuries can be significantly reduced by having a clear understanding of the underlying theories of fire combustion combined with a well thought out plan of action to accomplish the activities of salvage and overhaul. Understanding the interrelationship between prefire planning/ postfire conferences, proper ventilation, and good salvage and overhaul practices are vital steps to successful fire fighting and safe fire ground operations.

KEY TERMS

Churning of the air   The phenomenon of smoke being blown out at the top of the building opening, only to be drawn back into the structure at the bottom of the opening by the slight negative pressure (vacuum) created by the action of the mechanical fan.

Control valves   Valves used to control the flow of water in sprinkler or standpipe systems.

Fire management zone (FMZ)   A zone within a jurisdictional engine company's area where similar hazards are grouped by approximately equal needed fire flow and hazard. Once grouped, they receive a hazard severity ranking. Manpower and equipment response can be established using the hazard severity ranking given to each fire management zone.

Live load   Items inside a structure that are not attached or of a permanent nature.

Louvering   A method of roof ventilation that reduces the exposure to smoke and initially prevents cutting the supporting roof joists. It is especially effective on plywood-covered roofs.

Negative pressure ventilation   A method of forced ventilation that pulls air and smoke out of a structure.

Overhaul   A systematic process of searching the fire scene for possible hidden fires or sparks that may rekindle; also used to assist in determining the origin and cause of the fire.

Pike pole (rubbish hook)   A device with a sharp pointed head and a curved hook which is used to thrust into the ceiling area and then using the hook pull the ceiling materials downward. Because the hook is attached to a long pole, firefighters use it to ensure that all materials are removed when venting a roof from topside. The pike pole can also be used to sound the roof.

Positive pressure ventilation   A process that uses mechanical fans to blow air into a structure to remove smoke and gases.

Postfire conference/Postincident size-up   Following the incident, this analysis is conducted to exchange what was observed, the result of actions taken, and what lessons were learned so improvements can be made on future incidents.

Preincident planning   A process of preparing for operations at the scene of a given hazard or occupancy.

Salvage   Described as the protection of property which was not damaged by the fire or could be damaged by the fire, smoke, water, and related extinguishment activities.

Scuttle hole   An opening that allows entry into the attic area. The cover during ventilation in some cases can be removed to ventilate the attic below.

Siamese connection A hose fitting designed with a double hose line connection that allows two hose lines to be connected into one (generally larger) line.

Smoke ejectors Mechanical smoke fan designed to draw heated air and smoke outside of a structure.

Sounding a roof Using an axe or pike pole, firefighters search the roof ahead of them to determine if it is solid enough to handle the additional weight of firefighters.

Thermal imaging device A device that distinguishes between objects or areas with different temperatures and displays hot areas and cold areas as different colors on a video screen.

Ventilation A systematic process to enhance the removal of smoke and fire by-products to allow the entry of cooler air and facilitate rescue and fire fighting operations.

REVIEW QUESTIONS

1. Relate prefire and postfire planning activities to the need to learn and understand the fire combustion processes.

2. Describe the importance of preplanning activities, including safety concerns.

3. Why is it important during a postfire conference for firefighters to participate in a postincident size-up?

4. When cutting ventilation holes, it is sometimes necessary to make louver cuts. Describe a louver cut and include in your discussion when and why it is used.

5. Explain some fire combustion concepts that are important when conducting salvage and overhaul operations.

REFERENCES

Angle, James et al. (2008). *Firefighting Strategies and Tactics* (2d ed.) (Clifton Park, NY: Delmar Cengage Learning).

Delmar Cengage Learning. (2008). *Firefighting Handbook: Firefighting and Emergency Response* (3d ed.) (Clifton Park, NY: Delmar Cengage Learning).

Diamantes, David. (2007). *Fire Prevention: Inspection and Code Enforcement* (3d ed.) (Clifton Park, NY: Delmar Cengage Learning).

Federal Emergency Management Agency. (2003). *Trends in Firefighter Fatalities, Due to Structural Collapse, 1979–2002,* FEMA and NIST Report NISTIR 7069.

Federal Emergency Management Agency. (2003, March). *Rapid Intervention Teams and How to Avoid Needing Them,* FEMA TR-123.

Gagnon, Robert M. (2008). *Design of Special Hazard & Fire Alarm Systems* (2d ed.) (Clifton Park, NY: Delmar Cengage Learning).

Klinoff, Robert. (2007). *Introduction to Fire Protection* (3d ed.) (Clifton Park, NY: Delmar Cengage Learning).

Mittendorf, John H. (1996). "Strip Ventilation Tactics," *Fire Engineering* 149(3).

Smoke, Clinton H. (2005). *Company Officer* (2d ed.) (Clifton Park, NY: Delmar Cengage Learning).

Chapter 7

HIGH-RISE BUILDING FIRES

Learning Objectives

Upon completion of this chapter, you should be able to:

- Understand and explain why high-rise buildings present a difficult and different fire problem for firefighters, including the unique fire behavior problems that may be encountered in a high-rise fire.

- Recognize the difference in construction methods of high-rise buildings and explain how different construction materials and designs impact fire behavior in these buildings.

- Describe the fire fighting strategies and tactics used to locate, confine, and extinguish high-rise fires.

- Describe the special problems that may be encountered on high-rise fires such as communications issues, the stack effect, ventilation concerns, evacuation issues, and elevator control.

- Describe and explain the purpose of the special fire protection equipment which may be found in high-rise buildings.

- Describe when a stairwell support system may be needed.

INTRODUCTION

Development of vacant land in large cities has resulted in a tremendous increase in property values. As the amount of available land becomes scarce, the cost of the land increases, resulting in building developers building upward, rather than purchasing additional land to build horizontally. This upward growth has resulted in what firefighters call high-rise buildings, which are defined by the National Fire Protection Association (NFPA) standard 101 *Life Safety Code* as, "A building more than 75 ft (22.5 m) in height where the building is measured from the lowest level fire department access to the floor of the highest occupiable story."

There are variations in the code definitions of high-rise buildings throughout the country. For firefighters, the best policy is to check locally for the appropriate code to obtain a better understanding of the code specifics. Almost all of the definitions indicate that these buildings have floors higher than the reach of fire department ground ladders. This requires fighting the fire from inside the building. Upper floor fires present firefighters with a different set of problems than those encountered in buildings that can be easily and quickly reached from the ground level.

While the basic fire fighting tactics used on high-rise fires are the same as those used on most other structure fires, these buildings present special considerations and problems.

- The building may contain multiple occupancies such as apartments, commercial stores, restaurants, and extensive parking facilities.
- The height of the building presents the opportunity for greater smoke movement within the building.
- All fire fighting equipment, supplies, and human resources must be delivered to the upper floors, a labor-intensive task that requires additional personnel.
- Access to and egress from high-rises can be difficult under emergency conditions as occupants and firefighters are required during some emergencies to use the same stairwells to exit and enter the building.
- The HVAC systems can, if not properly controlled, assist in the spread of fire and smoke into other locations in the building.

Because of these significant differences, a description of the fire problems and fire behavior of high-rise buildings is of special importance to firefighters.

HIGH-RISE BUILDINGS

According to the National Fire Incident Reporting System (NFIRS), for the period from 1996 to 1998, there were 15,500 high-rise fires in the United States. These fires resulted in $252.3 million in annual property loss, and 930 civilian injuries including 60 civilian deaths. While 75% of these fires were in high-rise residential properties, they caused only 25% of the total fire dollar losses. The main reason for this is that high-rise residential fires tend to be contained within one room, whereas 75% of fires in high-rise manufacturing, industrial, and storage buildings are not contained to the room of origin. Figure 7-1 illustrates the dangers that will be encountered when faced with a fire in a high-rise.

Occupants of high-rise buildings face several important issues in terms of fire behavior and safety. These issues include the manner and materials of construction as well as the building contents. Firefighters categorize the construction materials and design of high-rises into three major time periods. Each time period has a distinctive construction design and use of construction materials.

Figure 7-1 *Fire in high-rise buildings can present firefighters with unique challenges.*

Early Fire-resistive Buildings (1870 to 1920)

Structures built from 1870 to 1920 were constructed with little or no concern for fire safety. At the time there were no standards for the protection against heat for the building's steel components. When heated a little above 1,000°F (537.8°C), steel quickly loses it strength. Steel also expands at a rate of 0.06% to 0.07% for each 100°F (37.8°C). This means that at 1,000°F (537.8°C), it will expand about 9% for each 100 ft in length. In a masonry building where the steel beam is blocked by a wall, the steel can cause a partial or total wall collapse.

Some of these early buildings were built with the floors supported by concrete piers which created an open void space where fire and heated gases could move underneath the structure. To prevent fires from spreading from these open spaces, some building contractors used terra cotta tile to provide fire resistance. These tiles were effective as long as they did not contain holes or poorly cemented joints, which would create open spaces and allow fire to spread. Most construction work used the least expensive materials and the workmanship was of poor quality resulting in a fire which would quickly spread throughout the entire building.

Some of these buildings were equipped with standpipes but most water supply line installations were too small to effectively deliver the amount of water needed during a fire. Preincident planning is recommended for jurisdictions that have older buildings remaining. Careful planning and a **risk assessment** should be done as well. Figure 7-2 shows a pre-1920 high-rise.

Risk assessment
Activities that involve the evaluation or comparison of risks and the development of approaches that can change the probability or consequences of an incident.

Figure 7-2 *A high-rise building constructed prior to 1920.*

High-rise Construction (1920 to 1960)

Buildings constructed from 1920 to 1960 have important fire safety features that were lacking in earlier buildings and have some features that are not found in the newer high-rise buildings.

These buildings were steel framed and tiled with concrete or masonry which was used for fire-resistive protection. Concrete steel reinforced columns that were spaced at about 30 ft apart supported the floors. Each floor was served by several widely spaced concrete encased steel stairwells. The windows on each floor could be opened, allowing firefighters to ventilate. Many of the windows leaked air so air movement was within the immediate floor area rather than throughout the entire building.

A good example of this type of building is the Empire State Building, as seen in Figure 7-3, which was built in 1930. It has thick masonry walls resulting in the building weighing about 23 pounds per cubic foot while newer high-rise buildings using drywall average approximately 7 to 8 pounds per cubic foot. The added mass will absorb more heat from a fire than the lighter-built building.

Figure 7-3 *Empire State Building.* Photo Courtesy of Michael Slonecker.

The nonabsorbed heat is reflected back into or contained within the room area, resulting in a hotter interior environment for firefighters.

High-rise Buildings Constructed after 1960

Buildings constructed after 1960 are the modern steel and glass buildings that can be found in most downtown areas of urban cities. These buildings differ from the earlier high-rise buildings because of new construction materials, methods, and systems. They use gypsum wallboard and lightweight concrete for fire-resistive protection. They also contain central core construction and use central heating units, ventilation, and air-conditioning systems for heating and cooling.

Gypsum board
Used in interior finishes for walls and ceilings.

Gypsum Wallboard/Plaster Board The product called **gypsum board** has gypsum material pressed into sheets 4 ft by 8 ft or 4 ft by 12 ft in size. The gypsum is held in place by two pieces of heavy paper to give it shape. The board can be nailed to wood studs or fastened with screws into metal studs. It is relatively light in weight when compared to other concrete products, but its strength quickly fades when the material is wet or suddenly hit.

On 9/11, in the World Trade Center, sheets of wallboard were knocked off the stairwell walls by the impact of the aircraft. This event caused the displaced sheetrock to block the stairwells, and trapped persons by blocking movement up and down the stairwells. The NIST study of this tragic event recommended that building egress systems should be designed to maximize the remoteness of the egress components, which would provide access and egress locations in more locations. Additionally, the construction materials used to protect these systems should provide a safe haven from fire activity inside the building itself.

Lightweight (blown-on) concrete
A lightweight concrete mixture that is blown onto metal to provide a coat of fire-resistive material.

Lightweight or Blown-on Concrete Firefighters use a coating of concrete called **lightweight** or **blown-on concrete** to help protect the steel structural components in fire-resistive properties. This type of concrete material contains additives to increase its heat resistance and adherence to the steel components. Its purpose is to insulate steel from the effects of fire and heat. The thickness and the ratio of concrete to trapped air determine the fire resistance. A wide variety of these materials are available, sold under various trademark names. Like gypsum board, after 9/11, questions were raised about the application processes and the poor adherence of the lightweight concrete that had been used to protect the steel floor truss components from fire and heat. New standards for application methods of the materials are being considered. Figure 7-4 illustrates a parking garage with steel columns and beams that is being protected by blown-on concrete.

Central Core Construction Also found in the post-1960s high-rises is the central core method of construction. In pre-1960 construction, exits could be found on every floor remotely located from each other and in most buildings one exit in every corner of the building. These exits served as a stairway to the first floor or the floor that contained the exit directly onto the street.

Figure 7-4 *Parking garage with steel columns and beams protected from fire with protective blown-on concrete.*

Figure 7-5
Plumbing, telephone, and electrical systems can be channeled within utility chases.

Central (or center) core construction
The building is constructed where the elevators, stairs, and support systems are located in the center of the building.

In contrast, the 1960 **central core construction** of buildings placed the elevators, stairs, restrooms, and room containing the building support systems in the center of the building. Thus, all the plumbing, telephone, and electrical systems could be channeled within utility chases (see Figure 7-5) centrally, reducing materials costs and installation costs as all the plumbing and wiring was centralized in one location.

One problem is that these vertical channels now provide an excellent method of travel for fire and toxic gases. Likewise, exit stairways can be constructed at less cost if some of the walls can be used as common walls for elevators and other exits. The idea is to centralize these common requirements in the center of the building leaving the outer portions of the building available for lease (Figure 7-6).

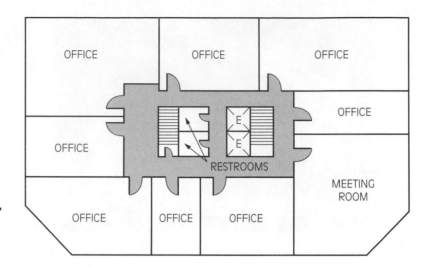

Figure 7-6 *Central core floor plan for high-rise building.*

HVAC Systems Since the late 1960s, architects and building owners have realized that energy that provides a comfortable environment is not being used efficiently. It is costly to draw air in from outside, filter out dust and impurities, heat it or cool it, and move it to the occupied areas. In effect, this air is a valuable commodity. If it can be kept within the building and recycled, the cost of operating the building can be substantially lowered.

With this concept in mind, architects seal these buildings. Windows no longer open and as an added efficiency feature, the air conditioning for several floors can be done from one central floor (mechanical equipment room).

This presents severe problems in the event of a fire. First, windows will have to be broken in order to provide horizontal ventilation, if it is required, and the HVAC system must be shut down during a fire or designed with features to protect the system from moving the fire from floor to floor.

In these systems, the circulating air is either heated or cooled and then forced by mechanical fans throughout the building or in higher buildings within a zone of the building. They use return ducts or a **plenum space** to return the air to the driving fan. On the air return upside of the system, the air is moved by negative pressure through the ducts or plenum to the heating or cooling element where it is cooled or heated and returned via the ducts again by positive pressure created by fans.

Fire dampers are installed in the return ducting to prevent fire and heated gases from being circulated in other parts of the buildings. Firefighters need to know the location of the fire dampers and to follow up to make sure fire and smoke is not being circulated or spread into other parts of the building when they are activated. After the fire has been extinguished, firefighters need to ensure that the fire dampers will be reset for future operation.

Only 15% of the total air volume is brought into the system per air circulation cycle, which reduces the total amount of heating or cooling needed. This is

Plenum space

A space above the suspended ceiling or hallway that is kept under negative pressure for return air.

a substantial cost savings over the older heating and cooling systems where none of the heated or cooled air was reused.

The systems in some buildings can be configured so that the total system or the zone within the building can be used to exhaust smoke from some areas and supply fresh air to an area in the event of a fire. This requires a thorough working knowledge of the specific building system and a detailed knowledge of how these systems function. On the other hand, some systems must be shut down immediately as they will spread fire and smoke into unaffected portions of the building. Information about the HVAC system should be discussed with the building maintenance service and written operating details should be obtained during prefire planning sessions.

STACK EFFECT

Stack effect
The temperature difference between the outside temperature of the building and the temperature inside the building.

The **stack effect** is the natural movement of air within a tall building caused by the temperature difference between the outside and inside of the building. This temperature difference causes the air in the building to rise or fall and varies with the amount of temperature difference. The air movement becomes very noticeable in buildings over 60 ft high and becomes stronger as the building gets taller and the temperature difference becomes greater.

For example, if the temperature outside a sixty-story office building is 100°F (37.8°C) and the temperature inside is 70°F (21.1°C), then there is a 30° difference. This difference will drive air current in the building in a downward movement. On the other hand, if it is 40°F (4.44°C) outside and 70°F (21.1°C) inside, the temperature difference is 30° again, but now the direction of the air movement in the building is upward.

Stratification location
The location in a high-rise building where light heated air flows upward and reaches a point where it is the same temperature and weight as the surrounding air.

This pressure difference forces heated air in a structure at lower levels up to the upper levels until this lighter heated air reaches a point in the building where the air temperature is equal and the weight of the air is balanced. This area in the building is called the neutral plane or **stratification location,** in which air flows upward and then flows horizontally. For firefighters, this means heated smoke and fire gases will begin to move horizontally resulting in lateral fire spread. In most cases, it occurs at or close to the middle of the building as the pressure difference balances out (see Figure 7-7).

Smoke and gaseous products of combustion can be influenced by this effect. In fire occurring in the lower levels (below the neutral plane), the internal pressure created by the stack effect draws the products of combustion toward any shafts or stairway openings. In fires occurring at higher levels (above the neutral plane), air movement draws products of combustion away from shafts or stairs toward the exterior of the building. In fires occurring within the neutral plane, the stack action has little effect on smoke movement. Figure 7-8 identifies how the temperature differences impact the stack effect.

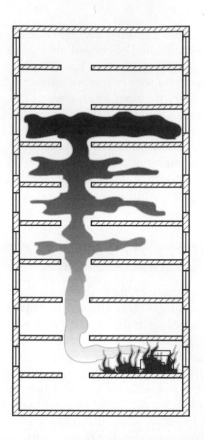

Figure 7-7
Stratification inside a high-rise building.

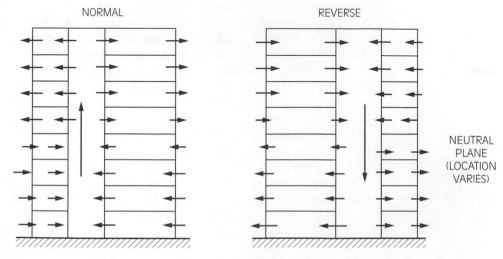

Figure 7-8
Temperature differences impact the stack effect in high-rise buildings.

AIR MOVEMENT DUE TO NORMAL AND REVERSE STACK EFFECT

NORMAL   REVERSE

NEUTRAL
PLANE
(LOCATION
VARIES)

NOTE: ARROWS INDICATE DIRECTION OF AIR MOVEMENT.

VENTILATION

Ventilation is the planned and systematic removal of smoke, heat, and gases from a structure. The reasons for ventilation are to remove toxic gases and to eliminate hazardous health conditions for occupants and firefighters. This will allow firefighters to locate the fire more quickly and reduce fire and smoke damage.

The traditional methods to accomplish ventilation at fires are opening or breaking windows, bulkheads, and scuttles or cutting a hole in the roof. These methods are not always available on high-rise building fires. In the older high-rise buildings equipped with windows that are opened for ventilation, it can be accomplished by opening windows on the fire floors and using positive or negative ventilation to force or pull the smoke and heated gases outside the building. In the newer high-rises (those built after 1960), ventilation is more complex as most buildings are sealed. Furthermore, these buildings built in the latter part of the 1960s and 1970s will probably not have an HVAC system configured to exhaust fire gases and smoke. Therefore, if horizontal ventilation is not an option (it is too dangerous to break exterior windows), then an enclosed stairwell can be used if it is not being used for occupant egress or firefighter access and if it contains an opening that can be used to safely exhaust smoke.

If a stairwell is available, then the door on the fire floor into the stairwell is to be held open and an exhaust fan is used to direct the smoke up the stairwell and through the enclosed shaft and out of the building in a safe location. All intervening floors not impacted by fire must be closed and all other openings should be closed so that fire and smoke will not spread to other areas of the building. This operation must be carefully planned and checked continuously for unwanted fire and smoke spread during the process.

Some newer buildings are equipped with fans that create a positive air pressure in the stairwell. These systems can be used but with great care and caution as the pressurized air can move smoke and fire into uninvolved portions of the structure. This is an important feature to learn more about from building engineers and maintenance personnel during preincident planning visits.

ELEVATORS

A key component of efficient high-rise fire operations is the safe and reliable use of elevators to transport firefighters to and from the fire location in the building. However, we know from past experience that elevators have malfunctioned during emergency operations resulting in the death of civilians, building security personnel, and firefighters. Generally, firefighters will attack a high-rise fire using the stairwell on fires occurring on any one of the first six floors. Fires on the lower floors do not warrant the use of an elevator unless its safety can be verified by sending a firefighter to the fire floor, where the elevator can be inspected for safety concerns.

When the incident is higher than the sixth floor, most departmental polices allow the elevator system to be used only after confirmation of the elevator's safety. A recent report by the National Institute of Standards and Technology (NIST, 2004) concluded that after traversing more than twenty floors, firefighters were so fatigued they were not prepared for fire fighting duties once they arrived on the fire floor. This finding confirms the importance of developing methods and procedures for firefighters to safely use elevators.

Some departments allow elevators remotely located from the fire to be used if there is a fire separation between the elevators and the fire area. Firefighters using these elevators move the personnel and equipment to the floor below the fire, then ascend the stairway one floor and travel horizontally to the fire. This approach is much safer and easier than ascending multiple stairs and also keeps firefighters that are loaded down with heavy, bulky equipment out of the stairwells where civilians may be descending to exit the building.

An understanding of local fire department policies and procedures is important when regarding if and when elevator use in high-rise firefighting is safe and will be permitted.

Use of Elevator Systems

Recent firefighter deaths and injuries are responsible for laws being passed requiring the installation and retrofitting of fire service override systems on elevators in newer high-rise buildings. These systems provide manual control over all automatic elevator controls in the cars, providing a greater margin of safety for firefighters and building occupants. The standards for fire service and elevator safety are in the American Society of Mechanical Engineers (ASME) code. They provide for uniform standards throughout the nation for newly constructed systems during installation and when older elevators are upgraded.

By installing the firefighter's key into the system, the system is taken out of automatic operation and placed under the manual control of firefighters. The manual control system is divided into three phases.

Phase 1 is sometimes referred to as the firefighter's recall. It brings the cars nonstop to the lobby or predesignated area and automatically opens the doors. In some systems, the phase 1 operation can be started by the activation of a smoke detector system in the building. The purpose is to ensure that the building occupants use the stairways and to prevent the car from stopping on the fire floor. Finally, it does not allow the doors to open automatically.

(*continued*)

Phase 2 is the operation of the fire service controls in the elevator car. This phase allows firefighters to manually control the elevator from inside the car. The lobby key in phase 1 cannot override the key operation in the car. Many departments have a standing policy prohibiting the use of any elevator car that does not have phase 2 fire service controls.

Phase 3 is connected with the building fire/smoke alarm system and applies when a fire or smoke alarm is activated. If the smoke detector activates in the lobby, the elevator cars will not return there but will be recalled to a predesignated floor other than the lobby. Firefighters will have to go to the designated alternate floor to get into the car. The location of the car can be found by viewing the panel in the lobby. During the preincident planning walk-through, this should be checked and

Figure 7-9   *Elevator panel showing the fire department key control functions.*

(*continued*)

noted in the building preincident planning document. Figure 7-9 shows an elevator panel with fire department control functions.

Following are general safety rules for use of elevators.

- Use the elevator key to place and keep all elevator cars in multiple hoist ways under fire department control.

- Never use an elevator that does not feel or look safe, and when using the system do not overcrowd the cars with firefighters.

- Never take the elevator directly to the fire floor. Always stop two or more floors below the fire and then walk to the fire floor. Department policies often dictate on which floor to exit the elevator.

- It is recommended to stop the ascending elevator every fifth floor to check for smoke. This is done by opening the elevator escape hatch and checking with a flashlight for smoke.

- There are several dangers when using elevators; the first is that the doors may open on the fire floor allowing heated fire gases to enter the elevator car trapping firefighters before they have the protection of hose lines in place.

- Always wear protective clothing, including SCBA, forcible entry tools, and a safety line.

On some fires, firefighters have been trapped in elevators that stopped in between floors as the power was interrupted. Firefighters entering elevators must always bring tools, ropes, and extra air bottles with them not only as backup for the fire fight, but also in case they become trapped in an elevator.

STAIRWELLS

In most high-rise buildings, the occupants are generally mobile and are able to use the stairwell to exit the building on their own. However, the congestion of firefighters moving upward in the building while occupants are attempting to evacuate reduces not only the firefighters' movement but also inhibits the escape of the occupants. One way to avoid this problem is to designate one or more stairwells to occupant egress and one or more to fire fighting operations.

Stairwell support procedure

A procedure used to move resources to the fire attack area when the elevators are unsafe for use.

Elevators should be used only if they are safe; if not, a **stairwell support procedure** must be put in place. Using this procedure, firefighter teams are stationed two floors apart in the stairwell. The first team carries the equipment up two floors to waiting firefighters, who then carry the equipment two more floors. In the meantime, the team descends and rests while waiting to carry and pass on additional equipment. While this method requires a large number of firefighters, it is an option available to supply needed resources to the fire attack area.

Pressurized Stairwells

Newer high-rise buildings have enclosed pressurized stairwells designed to provide the occupants and firefighters a smoke-free environment while moving within the stairwell. The basic principle of a pressurized stairwell is to enclose the stairwell tightly and then add a fan to increase the air pressure inside the stairwell so when the door or doors to enter the various floors are opened, the air pressure inside the stairwell will push the smoke back and keep the smoke in the floor area. In some systems, if too many doors open at the same time, the pressure reduces to a point that the stairwell fills with smoke. Fire department fans can supplement the pressure in the stairwell if needed.

BUILDING FIRE PROTECTION SYSTEMS

The important tactical consideration with fires in high-rises is the reliance of fire departments on built-in systems to assist in fire control and extinguishment. Building systems available to firefighters are:

- standpipes;
- sprinkler and water supply systems;
- fire pumps;
- fire communications and command systems; and
- pressurized stairwells.

These systems are varied and can be complex. Fire departments must master their use in order to bring fire operations to a successful conclusion. A closer look during fire preincident planning sessions at some of the systems found in high-rise office buildings is needed so these systems can be employed properly during fire operations.

Standpipes

Most building codes require standpipe systems in high-rise buildings. Although some buildings are not technically considered high-rise buildings, they should be defined as such for firefighting purposes, as the fire service cannot reach the interior of the building from ladders. As discussed earlier, this definition would include buildings that have stories beyond the reach of available ladders.

One way to better understand standpipes is to look at them as fire water mains that are built into the building with an outlet. These fire hydrants are located in the stairwell of every floor. Firefighters connect into the hydrant and then advance the hose line into the fire floor from the stairwell. On the outside of the building, the standpipe connection is where the fire pumper connects to pressurize the water main. Preincident training exercises are a good way to

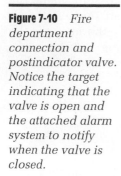

Figure 7-10 *Fire department connection and postindicator valve. Notice the target indicating that the valve is open and the attached alarm system to notify when the valve is closed.*

determine if a building has a workable standpipe system. Figure 7-10 shows a fire department connection and postindicator valve.

Because the fire pumper pumps one set pressure, the pressure at each floor will vary and in some cases can be too high for firefighters to control. Therefore, in some cases, a pressure reducer will need to be installed. Firefighters must ensure that their operation and the pressure is correctly set during the preincident planning session.

Sprinklers and Water Supply Systems

Most new high-rise buildings are equipped with fire sprinklers. Because of the height of these buildings, many of the systems are installed in the high-rise in zones so that fire pumps can provide sufficient water pressure.

When properly installed and maintained, fire sprinkler systems can quickly control and extinguish most fires with water from two or less heads. In 1991, the One Meridian Plaza office high-rise in Philadelphia, Pennsylvania, was undergoing sprinkler retrofitting but was not yet completed. A fire started on the twenty-second floor and burned to the twenty-ninth floor (the floors without sprinklers). On the thirtieth floor, ten sprinkler heads extinguished the fire. The setting of the standpipe pressure regulator valves also impacted this incident.

In taller high-rise buildings, the building is divided into zones with a water supply tank installed inside each zone. The tank is then filled from outside water supplies using a pump or a series of pumps to fill the tank. During a fire, the tank and a fire pump are used to supply water for the fire sprinkler system and standpipes within that zone. Firefighters need to be familiar with the operation of these water tanks and pumps and the location and coverage of the various zones in the high-rise building.

Fire Pumps

Fire pump

A specially designed and listed pump that increases the pressure of the water serving a fire protection system.

All high-rise buildings are required to have auxiliary **fire pumps** designed to increase the pressure in the sprinkler and/or standpipe system. These pumps can be found singly or in pairs. They are designed to start when the pressure drops; then a sprinkler head or standpipe valve is opened, decreasing the pressure in the pipe. The pressure decrease causes a switch on the pump to note the drop in pressure and starts the pump.

Because most fire sprinklers and valves are rated at 175 psi, a pump relief valve might be necessary to prevent overpressurization that could damage sprinkler components when pump discharge pressures exceed 175 psi. Pressure regulating valves are also permitted to be installed on the discharge side of the fire pump to protect sprinkler components in high-rise buildings where the pressure required at the base of the riser exceeds the rating for the system components on the lower levels.

Although some systems may be more complex than others, the firefighter should know how to operate it and have a basic understanding of fire pumps. It is essential that the inlet and discharge gauges be checked and it is critical that the bypass valve is left closed after pump testing. Fifty percent of the pump capacity can be lost with an open bypass valve.

Generally, fire pumps are activated by following certain procedures that should be posted on the control panel. Once the firefighter starts the pumps, he or she should remain nearby to ensure that the required pressure is maintained and that no one stops the pumping operation without permission from the incident commander.

Fire Communications Systems

In most postfire reviews for high-rise building fires, the issue of poor fire ground communications is often raised. The reason is that high-rise buildings are built with steel columns and beams enclosed with concrete which absorbs much of the energy from fire department radio communications systems. In many cases, this steel and concrete enclosure makes emergency communications totally ineffective.

To resolve this problem, newer high-rise buildings have communications systems built into the structure. These installed communications systems use hard wiring that allows intrabuilding communications using a series of high-reliability speakers and telephone devices located throughout the building. This system allows the fire department to direct occupants during evacuation procedures and allows the incident commander to speak directly to the operations and division chiefs.

SEARCH AND RESCUE

The key to successful search and rescue efforts in high-rise buildings is the same as other buildings. Establish a systematic search method to ensure all areas are checked for remaining occupants and set up an accountability system to account for those found and those missing.

Convergence cluster behavior
Persons who feel threatened gather as a group to gain a feeling of safety in the presence of others.

Firefighters need to be aware of a newly found behavior called **convergence cluster behavior,** which was first observed during the 1989 Las Vegas MGM Grand hotel fire. Convergence cluster behavior occurs when occupants feeling threatened by fire converge together to feel safer in the presence of others. This results in a group in one area or room rather than people waiting for rescue in their separate rooms. On some fires, firefighters have wasted time and effort searching vacant rooms only to find all persons gathered in one location. An accountability system is needed to verify the whereabouts of all occupants.

Relocation of Occupants

Total evacuation of a high-rise structure during fire fighting operations is generally neither practical nor feasible. The primary effort depends on the number of people needing to be moved, how they can be moved, and if a safe refuge area is available within the structure. An example of this is providing safe areas or zones in hospitals, as patients can be moved horizontally to another safe zone within the building.

■ Note
During 9/11, sheet rock and areas of safe refuge did not work well in providing protection for the building occupants. Efforts are underway to find methods to strengthen building construction in safe refuge areas. Hi-impact© wallboard is now available and may be suitable but will need to be tested further.

SALVAGE

High-rise office buildings, hotels, and apartments typically have valuable contents, such as computers, office equipment, business records, and personal items. The owners and occupants may request assistance in obtaining these items. When and if possible, it is important for good community relations to make every attempt to assist these individuals.

The height of the building comes into play in an opposite way when we consider property conservation. In life safety and fire extinguishment, people and property above the fire are normally considered to be the most at risk. Fire, smoke, and heat will first affect materials on the fire floor and the floors above. However, the greatest need for property protection is generally downward as water flows through stairs, walls, utility chases, electrical fixtures, and other openings, damaging and destroying more property beneath the fire. In addition to protecting valuables with covers, firefighters should channel the flow of water down the stairs or through drains. Like life safety and fire ground operations, property conservation operations must be well planned and systematically executed to reduce loss.

OVERHAUL

Overhaul of high-rise buildings is labor intensive. A good preincident plan helps to determine where to direct initial overhaul efforts. The preincident plan can direct fire fighting forces to hidden shafts that need to be thoroughly checked as well as false ceilings which should either be pulled down or opened so areas above them can be checked for fire extension. Crews should be assigned to every floor above the fire to check for smoke and possible extension. Additionally, crews should check areas below the fire to verify complete extinguishment.

SUMMARY

High-rise fires in the United States are becoming an increasing and significant fire and life safety problem. Looking at the U.S. statistics for fatal high-rise fires in offices, hotels, and apartments, there were more fatal fires annually in apartment high-rises than in office and hotel high-rises. The hotel and office building fires get the most attention because large numbers of people are often killed in a single incident.

There are a number of significant differences between ordinary structure fires and high-rise structures that make fighting these fires difficult. Almost universally the communications inside them is poor, the height of the building inhibits using ground ladders for easy entry into the upper portions of the building, and the use of elevators is dangerous if the proper safety procedures are not used. Ventilation of enclosed high-rise buildings is another problem, when the movement of hot fire gases can be spread into areas not affected by the original fire if ventilation is not properly conducted.

Firefighters need to be aware of and know how to operate the special equipment found in high-rise buildings. Special attention needs to be given to determining if convergence cluster behavior is occurring and if problems related to evacuation can be resolved by relocation. Salvage and overhaul operations will need special attention and careful planning to reduce smoke and water losses.

KEY TERMS

Central (or center) core construction   The building is constructed where the elevators, stairs, and support systems are located in the center of the building.

Convergence cluster behavior   Persons who feel threatened gather as a group to gain a feeling of safety in the presence of others.

Fire pump   A specially designed and listed pump that increases the pressure of the water serving a fire protection system.

Gypsum board   Used in interior finishes for walls and ceilings. It is made with a core of calcinated gypsum, starch, water, and other additives that are sandwiched between two heavy paper sheets. It is also known as drywall or plasterboard.

Lightweight (blown-on) concrete   A lightweight concrete mixture that is blown onto metal to provide a coat of fire-resistive material. The fire-resistive rating varies with the material and its thickness.

Plenum space A space above the suspended ceiling or hallway that is kept under negative pressure for return air.

Risk assessment Activities that involve the evaluation or comparison of risks and the development of approaches that can change the probability or consequences of an incident.

Stack effect The temperature difference between the outside temperature of the building and the temperature inside the building. This temperature difference creates air movement inside the building, which may carry heated fire gases to the upper or lower floors.

Stairwell support procedure A procedure used to move resources to the fire attack area when the elevators are unsafe for use. Firefighter teams position themselves two floors apart in the stairwell to carry resources two floors and return to rest and carry additional resources again. This relay system provides a steady flow of resources to the fire area.

Stratification location The location in a high-rise building where light heated air flows upward and reaches a point where it is the same temperature and weight as the surrounding air; at this location it flows horizontally.

REVIEW QUESTIONS

1. Explain the reasons for the increase in high-rise buildings in America.

2. Provide your own definition of a high-rise building in your jurisdiction. If you are not in the fire service, describe the NFPA definition of a high-rise building.

3. Briefly point out some of the fire behavior problems you might expect to encounter during a high-rise fire.

4. Describe how more mass or more concrete in a high rise-building impacts fire fighting and fire behavior.

5. Describe the stack effect and its impact on the movement of fire and heated fire gases in a high-rise building.

REFERENCES

Angle, James S. et al. (2008). *Fire Fighting Strategies and Tactics* (2d ed.) (Clifton Park, NY: Delmar Cengage Learning).

FEMA. (2002). "High-rise Fires," *Topical Fire Research Series* 2(18), http://www.usfa.dhs.gov/.

Hall, John R. (1997). *High-rise Building Fires* (Quincy, MA: NFPA).

National Construction Safety Team. (2005). *Final Report of the National Construction Safety Team on the Collapses of the World Trade Center Towers* (Gainsburg, MD: National Institute of Standards and Technology).

National Fire Protection Association, Standard 101 *Life Safety Code*.

National Institute of Standards and Technology. (2004). NCSTAR 1.

Norman, John. (2005). *Fire Officer's Handbook of Tactics, Fire Engineering* (3d ed.) (Saddle Brook, NJ: Penwell Publishing Company).

Smoke, Clinton. (2005). *Company Officer* (2nd ed.) (Clifton Park, NY: Delmar Cengage Learning).

U.S. Fire Administration. (1991). *High-rise Office Building Fire, One Meridian Plaza, Philadelphia, Pennsylvania,* Technical Report 49, http://www.usfa.fema.gov/applications/publications/display.cfm.

———. (1993). *New York City Bank Building Fire, Compartmentation vs. Sprinklers, New York* Report 71, http://www. usfa.fema.gov/applications/publications/display.cfm.

———. (2002, June). *Multiple Fatality High-rise Condominium Fire, Clearwater, Florida,* Report 148, http://www.usfa.fema.gov/applications/publications/display.cfm.

———. (2003). *Risk Management—Planning for Hazardous Materials: What It Means for Fire Service Planning,* Report 124, http://www.usfa.fema.gov/applications/publications/ display.cfm.

Chapter

8

WILDLAND FIRES

Learning Objectives

Upon completion of this chapter, you should be able to:

- Explain the basic fire combustion principles and be able to apply them to wildland fires, and differentiate wildland fire behavior from structural fire behavior.
- Examine how weather conditions impact wildland fuels and the behavior of wildland fires.
- Describe the various parts of a wildland fire and identify how fire behavior impacts the methods of fire fighting wildland fires. Include some of the special techniques needed to extinguish and control these fires.
- Describe the method used to classify resources used on wildland fires and how fire behavior impacts the type and amount of resources needed to suppress wildland fires.
- Describe the various resources and tools used in extinguishment of wildland fires.

INTRODUCTION

There is a vast difference between wildland fire fighting and structural fire fighting. Careful study of these differences indicates variations in certain factors that impact fire behavior; however, all fires follow natural laws of physics and once learned they can be applied in both fire fighting arenas. Discussion in this chapter centers on the differences and similarities between both types of fire fighting, and how fire extinguishment methods differ as well.

This chapter begins with an overview of the basic components of combustion. The fire tetrahedron (described in Chapter 3) is the new model used to depict fire processes. Despite it being a valid model for fire extinguishment, the former triangle model is still used by most wildland agencies to explain combustion and extinguishment processes occurring on wildland fires.

The Fire Triangle for Wildland Fires

In wildland fire fighting the heat is cooled, the fuel is removed, or the oxygen is excluded from the fuel to stop the process of combustion. These actions are carried out by hose lines directing water at the heated fuel, or by hand tools or machines, which remove the fuel from the path of the heated fuel. Another technique is to use dirt or foam to smother, cool, and/or exclude the oxygen. Foam and other chemical applications can aid to accomplish extinguishment. Figure 8-1 illustrates the fire triangle.

Heat Removal

One side of the fire triangle is the heat side that represents its removal or reduction of the ignition temperature of the fuel. Once the fuel is cooled below the temperature at which it gives off vapors, combustion will no longer take place. In other words, the fire is extinguished. Water is the most effective fire-suppressing agent that is readily available. The key to using water is to use

Figure 8-1 *The fire triangle shows the wildland fire combustion process.*

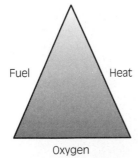

only the amount it takes to do the job. To do so, firefighters use a fine spray, breaking up the mass of the water so the maximum amount of heat will be absorbed by as little water as possible. The conversion of water droplets into steam allows the water to absorb greater amounts of heat energy. For deep-seated fires in forest materials, a straight stream is needed to penetrate and cool the material below its ignition temperature. For fuel giving off large amounts of radiant heat, a straight stream may be needed to supply sufficient water to reach the fuel.

On many wildland fires water is not available in abundance. Therefore, other extinguishment means need to be used. For example, small amounts of water from a hose line or back pump are used to cool the fire so firefighters can get close to the fire edge then use tools to remove the burning fuel from the path of the fire. This action breaks the fuel side of the triangle.

In the 1988 Yellowstone National Park fire, a compressed air foam system (CAFS) was used successfully in coating and protecting dwellings with Class A foam. Compressed air foam systems inject air into the foam solution with an air compressor. The foam is made by mixing 0.1% to 0.3% foam solution with 10% water and compressed air. This method significantly reduces the amount of water needed, thus enhancing its application to wildland fires. Figure 8-2 shows a firefighter applying Class A foam to a wildland fire. Because a large amount of water has been replaced by air in the hose lines, firefighters find the lines are easier to move.

In addition, the foam can be made wetter if needed, for use on a hot, advancing fire or drier to adhere to the vegetation. The foam provides a protective blanket from the fuel. The added detergent in compressed air foam allows water to break the surface tension and penetrate fuels easier and more quickly. Table 8-1 identifies the relationship between foam expansion rate and drain time. This principle of

Figure 8-2 *Class A foam applied on a wildland fire.*

Table 8-1 *Foam expansion rate and drain time.*

Expansion Ratio	Foam Type	Drain Time
1:1	Foam Solution	Rapid
↑	• Mostly water	↑
	• Clear to milky in color	
	• Lacks a bubble structure	
	Wet Foam	
	• Watery	
	• Lacks body	
	• Large to small bubbles	
	• Fast drain time	
	Fluid Foam	
	• Flows easily	
	• Consistency similar to watery shaving cream	
	• Medium to small bubbles	
	• Moderate drain time	
	Dry Foam	
	• Mostly air	
	• Looks like shaving cream	
	• Clings to vertical surfaces	
↓	• Medium to small bubbles	↓
20:1	• Slow drain time	Slow

Fire retardant

Any substance that by chemical or physical action reduces flammability of combustibles.

wetter (deeper penetration) to drier (lightweight protective covering) is illustrated in Figure 8-3.

Using the foam and **fire retardant** materials, firefighters can slow a large fire. These materials can be applied by aircraft and special foam application–equipped vehicles such as the CAFS.

Fuel Removal

In wildland fire suppression, removing fuel from the path of the fire is the most common method of extinguishment. These operations include cutting or scraping, using bulldozers and/or hand crews, or backfiring operations where fuel is burned in front of the main fire thus removing burnable fuel in the path of the main fire.

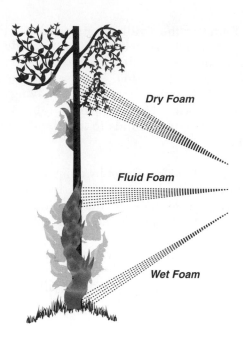

Figure 8-3 *Dry, fluid, and liquid foam applications.*

Dry Foam

Fluid Foam

Wet Foam

Oxygen Removal

Oxygen can be removed or excluded from the fuel on a wildland fire by throwing dirt directly at the base of the flames or at the area of combustion to cool down and exclude the oxygen from the fire. This action is followed by other extinguishing actions to ensure that the fire is completely out and cannot spread to any unburned fuel. If limited water is available this method reduces the total amount of water needed, but generally the penetrating ability of water is needed to extinguish any deep-seated fire.

Aircrafts are used on wildland fires to drop either water or flame retardant solutions that are designed to suppress the combustion process by cooling and smothering the fuel. To be approved for use, the flame retardant material must degenerate or be biodegradable generally within thirty days of application.

Wildland Heat Movement

Wildland fire heat output is transferred the same way as in structural fires. The heat moves in four ways: conduction, convection, radiation, and direct flame impingement.

Conduction Conduction is the transfer of heat within the material itself. Firefighters should be alert to the fact that fire can travel by means of metal fences or other metal objects found in the wildland areas. Although not conduction by the true definition, a fire in deep-seated ground litter underneath standing trees can transmit heat to the underground root system. In some cases, the roots will carry this fire across a prepared fire line. To detect and control this unseen fire, **cold trailing** is needed. Cold trailing consists of removing all fuel from a ground surface area (generally at least 18 inches in width) and then checking the ground underneath by using one's hands to feel the dirt and make sure underground fuels are not burning.

Convection   Convection is the transfer of heat by a liquid or a gas. In a wildland fire, the fire extends itself through heated air, which arises from the burning fuel. This movement of heat through the air transfers the heat into the cooler unburned fuel. If sufficient heat is present, it will raise (preheat) the fuel to its ignition temperature.

Radiation   Fire radiates heat waves (energy) ahead of the fire. In doing so, it preheats the fuel and in some cases increases the chances of **area ignition.** Once a fuel is preheated to its ignition temperature, all of the fuel in the area is ignited.

Direct Flame Impingement   Direct flame impingement occurs on wildland fires when the flame front is moving upslope where the flames *lay down* or move at an angle to the slope so that the flame front burns directly into the exposed fuel. This direct flame impingement increases the burning rate as it not only preheats the fuel, but the flames also provide the direct source of ignition. Figure 8-4 illustrates how a wildland fire can move uphill.

Cold trailing
A method of controlling a partly dead fire by carefully inspecting and feeling with the hand to detect any fire by digging out every live ember and trenching any live edge.

Area ignition
The ignition of an accumulation of heated gases from a number of individual fires in an area either simultaneously or in quick succession.

Figure 8-4   *A wildland fire can move uphill quickly when the flames lean to an angle and heat the fuel by convection currents and direct flame*

WILDLAND FIRE SIZE-UP

When sizing up a wildland fire, the same five-step decision-making process as described in Chapter 5 is used, but the components of COAL T'WAS WEALTHS are not needed unless structures are involved or exposed. On wildland fires, the five-step decision-making process is applied to each of the following factors.

1. **Factors impacting life safety.** Life safety concerns are those factors that may lead to the death of civilians and/or firefighters.

2. **Factors impacting property safety.** Property safety is a concern because of the opportunity for rapid fire spread from a hot wildland fire into structures, which may lead to a conflagration.

3. **Factors that may harm the environment.** One factor that may harm the environment is the damage and loss of the soil after burning away the vegetation as the roots work to hold the soil in place. Without the roots to hold the soil in place, the vegetation is washed away during the rainy season.

   A second environmental issue is the toxic particulates and ash given off during the combustion of the wildland fuels. Some smoke and ash materials carry acids and other toxic materials, which can impact paint and persons who are sensitive to these types of materials.

4. **Factors that will harm wild life.** The loss of vegetation results in the loss of food supply for many of the forest creatures.

5. **The availability of needed fire fighting resources.** The nonavailability of aircrafts and other resources can seriously hamper the efforts of firefighters as they attempt to control a fire in its early stages. The L.A. County Fire report (1989) found that 97% of wildland fires can be controlled if proper and sufficient resources are available during the first twenty to thirty minutes of the ignition. The 3% of uncontrollable fires are pushed by the wind or spread into structures that were exposed by the brush.

Burning index
A number in an arithmetic scale determined from fuel moisture content, wind speed, and other selected factors that affect burning conditions and from which the ease of fire ignition and their probable behavior may be estimated.

BEHAVIOR OF WILDLAND FIRES

Fire behavior is also affected by weather, topography, and fuels. These three main factors are often referred to as the "fire behavior triangle." Each factor of wildland behavior has subcomponents, which also affect fire behavior. One method that firefighters and wildland management personnel use is the **burning index,** which is a number in a arithmetic scale determined from fuel moisture content, wind speed, and other selected factors that affect burning conditions and from which the ease of fire ignition and their probable behavior may be estimated.

Weather

Understanding weather characteristics is essential for the safety and effective extinguishment of wildland fires.

Temperature In basic fire chemistry, fuel temperature is a factor that is absolutely necessary to initiate and continue the fire combustion process. Temperature is one component of the weather that impacts the behavior of wildland fuels. The range of ignition temperature of woody fuels is between 400° and 700°F (204° and 371°C). Normally the woody fuels will burst into flame at approximately 540°F (282°C). However, the time required to bring the wood to its ignition temperature will vary with the amount of moisture in the fuel as the heat must drive the water from the fuel in order to bring it up to its ignition temperature.

The highest temperature the sun can be expected to provide on a wind-sheltered slope surface is 150°F (66°C), which is far below the required temperature for spontaneous ignition. Nevertheless, the sun's heat plays an important role because it preheats the fuel, raising its temperature and thereby reducing the amount of outside heat needed to bring it to its ignition temperature. During the day the sun heats the fuel, but at night the air is cooler than the fuel so the heat is transferred from the fuel to the cooler air. Firefighters use this opportunity to complete fire extinguishment during this period.

Stable and Unstable Air Masses The relationship between the vertical temperature distribution within an air mass and a vertically moving parcel of air is termed **air stability and instability.** If air mass temperature decreases sharply with altitude, conditions are favorable for air currents to rise vertically through the air mass. This movement produces an unstable condition, which is dangerous, as is any movement of vertical and horizontal currents. At the same time, this creates an unpredictable wind situation.

Temperature Inversion A **temperature inversion** occurs when cooler air is found close to the ground surface and warmer air is above this pool of cooler air. Quite often this occurs along the coasts where a marine inversion is created when cool, moist ocean air flows inland and settles in low areas. If a fire in this low area heats the cooler low-lying air to a temperature that is greater than the warm air overlying the area, the heated air breaks through the **inversion layer** resulting in a quick rush of cooler air that causes the fire to accelerate and burn with much greater intensity.

Firefighters should be aware and concerned when a temperature inversion layer is present as the conditions can lead to a rapid change in the behavior of the fire. The inrush (entrainment) of a fresh supply of oxygen near the ground surface will increase the fire's intensity.

Relative Humidity Water vapor is simply water in a gaseous state and is an important ingredient in fire behavior. As air moves over the surface of water, it picks

---

Air stability and instability
The relationship between the vertical temperature distribution within an air mass and a vertically moving parcel of air.

Temperature inversion
An action that occurs when cooler air is found close to the ground surface and warmer air is above this pool of cooler air.

Inversion layer
A layer of comparatively warm air overlaying cool air. In this stable atmospheric condition, the atmosphere will resist vertical motion of the air.

Evaporation
The process by which molecules in a liquid state (e.g., water) spontaneously become gaseous (e.g., water vapor).

up water vapor. This process is called **evaporation.** The oceans are the primary source for adding water vapor to the atmosphere. Lakes, rivers, snow, and vegetation furnish lesser amounts.

Relative humidity is the ratio or the amount of water vapor in the air to the maximum amount that the air can hold. If relative humidity is 100%, the air is completely saturated; it can hold no more water. In contrast, if the relative humidity measures 50% then the air is only half saturated and is able to hold 50% more water vapor. Low relative humidity draws moisture from fuels allowing them to heat more rapidly. When the relative humidity is below 30% the fuel becomes very dry and makes conditions favorable for a wildland fire. If the reading is 15% or below, the condition is declared a **red flag alert.** During red flag alerts, fire departments prepare for wild fires by moving and adding additional resources for initial fire responses.

Red flag alert
The condition that occurs when relative humidity is below 15%, causing fuel to become very dry and thus making conditions favorable for a wildland fire.

Wind Velocity and Direction   Wind is the result of the uneven heating of the earth's surface. Heating the air causes it to become lighter; therefore, it rises, drawing in cooler air from below. Over the surface of the earth several major forces are at work unceasingly to stimulate air movement. There are vast areas of heating near the equator that are much closer to the sun where the heat produces rising air currents that return to earth in the cooler regions. This movement of warm to cooler air sets up a circulation pattern of rising and descending air currents. The pattern is impacted by the earth's revolutions, which deflects the rising air currents and causes them to move in a deflected horizontal direction.

The changing seasons alter the pattern of wind movement as well, because the hot and cold regions of the earth are shifted as the earth tilts in its relationship to the sun. This tilting results in further changes to the upward vertical, downward, and horizontal movements of the rising and descending currents. Firefighters need to prepare for the wildland fire season by attending training sessions, checking the local weather report daily, and holding preplanning sessions.

Foehn
A dry wind with a strong downward component locally called Santa Ana, North, Mono, Chinook, or East Wind.

Foehn Winds   Winds called **foehn** (pronounced "fane") are **gravity winds,** which occur when air is pushed over high elevations and flows downhill. Gravity winds also occur when a high-pressure system is located near mountain ranges where the airflow around the high-pressure system causes some of the air to spill over the higher elevations, resulting in a strong wind moving downhill at high speed. As the foehn wind flows downslope, it is compressed and becomes warmer and drier. This movement of warm air dries the fuels and reduces the fuel's moisture. During these conditions, winds can reach 70 mph or higher.

Although the foehn wind is responsible for a general drying influence on fuels and high rate of fire spread, the real hazard for firefighters comes from the day of transition. The day of transition is the first day when offshore winds subside and cool, and moist onshore flows begin gradually returning. On the

Gravity winds
Winds that occur when air is pushed over high elevations and flows downhill.

transition day, fire spread changes from downslope to upslope. This change may come at any time and, as a result, firefighters must always be on guard so as not to get trapped.

Some of the most costly fires in the western part of the United States occur when gravity or foehn winds push wild fires into populated **interface lands.** This is where structures are built close to or in the wildland area (defined as structures within 400 ft of the wildland area). These foehn winds have local names such as Santa Ana, Chinook, and Mono winds.

Interface lands
The areas where structures interface with wildland vegetation.

Fuel Moisture Generally, nighttime brings more moisture into the air as it has been cooled; this cooling effect from the moisture in the air (relative humidity) affects the amount of moisture within the fuel. Nighttime or early morning provides firefighters with the opportunity to take action when the fuel moisture is higher. At other times, if all other conditions are conducive for a wildland fire, fuel moisture does not make a difference.

Dust Devils and Fire Whirls Dust devils, mirages, and fire whirls are indicators of unstable surface conditions and will cause erratic fire behavior. These conditions are the result of uneven heating which causes ground temperatures to vary. Heated air rises and is pushed horizontally by the prevailing wind and in some cases it begins to whirl. The problem for firefighters is that these whirling winds throw burning materials ahead of the main fire front, creating **spot fires.** This causes fuel preheating and results in rapid fire spread.

Spot fires
Fires that jump over a control line and start the vegetation burning ahead of the main fire front.

Topography

Topography includes such elements as slope, aspect, elevation, and configuration or the lay of the land. Topography can be considered static, because the forces that change like rain and wind generally work very slowly. However, topography causes changes in fire behavior as the fire moves across the terrain. This erratic fire behavior is prompted by changes in the wind pattern caused by the slope, aspect, elevation, canyons, saddles, ridges, and chimneys found in the land surface.

Slope Slope affects the spread of fire in two ways. First, the fuel is preheated by direct flame impingement and convective air currents; and second, a draft effect is created. A fire will run faster uphill than it will downhill, as long as the wind is not strong enough to change the direction of spread. On the uphill side of the fire, the flames are closer to the fuel, causing preheating and faster ignition of the fuel. The heat rises along the slope causing a draft of the air and increasing the rate of spread. A rule of thumb is that for every 20% increase of slope, the rate of fire spread doubles. An important safety factor is to always check for fire below, as gravity can cause burning materials to roll downhill and cross control lines.

Bottom (toe)
The rear-most area of a fire, generally used as an anchor or starting point for suppression activities.

Aspect (exposure)
The direction a slope faces to the sun.

Flash fuels
Fuels such as grass, leaves, ferns, tree moss, and some types of slash which ignite readily and are consumed rapidly when dry.

Fuel type
An identifiable association between fuel elements of distinctive species, form, size, arrangement, or other characteristics that will cause a predictable rate of fire spread or difficulty of control under specified weather conditions.

■ Note
Special precautions must be taken when fighting in narrow canyons with intersecting canyons. The wind eddies that occur in these intersections can cause erratic fire behavior.

Fires that start at the **bottom** or **toe** of a slope generally become larger because they have greater amounts of fuel available to them and the preheating occurs on fires burning upslope. Preheating is a result of direct flame impingement, radiative heat, and convective air currents. The flame impingement, heat, and air currents make fires running upslope the common denominator of most fatal wildland fires.

Steep slopes also affect the ability of firefighters to work efficiently. The incline adds to the difficulty of constructing fire lines as well as moving hose lines. In addition, midslope fires present firefighters with great danger as the fire can move below them and suddenly erupt and move rapidly uphill trapping them.

Aspect   The direction a slope faces to the sun, called the **aspect** or **exposure,** impacts the spread of fire in several ways. South-facing slopes are exposed to more radiant heating by the sun, resulting in temperatures that are normally higher with lower humidity. The fuels are usually heated more and as a result are dry. The warming changes hourly as the sun moves across the sky, resulting in different fire behavior that is affected by the angle of the sun and its heating of the fuels. East-facing slopes start to heat up first with the rising sun.

Aspect also influences the type of natural vegetation. Fuels on a southwestern exposure generally present the greatest fire hazard because they are exposed to direct sunlight for longer periods of time. The fuels are smaller and drier. Fuel on the northern slopes receives more moisture and is heavier. Fires on the northern aspect generally move slower because of the moisture content of heavier fuel; however, once burning, these fuels generate more BTUs and as a result present a greater resistance to fire control.

Elevation   Fire behavior is affected by elevation as well. It influences how air moves from the valleys that are warming to the cooler ridges (warm air raises). Also, elevation can affect positioning of warm/cool air masses in thermal belts (the layering of air masses). It also affects the length of the fire season (the lower the elevation the longer the season), as the light **flash fuels** are drier and quick burning. Elevation also affects **fuel types** (vegetation types and fuel-loading change with higher elevation). Notice the grass to brush to timber change as the elevation increases in Figure 8-5.

Canyons, Saddles, and Ridges   The topography of the land affects wind patterns and fire spread. Canyons, saddles, and ridges directly influence how a fire burns. Wide canyons will have little effect on the wind pattern while narrow canyons may significantly impact the wind as it is driven down the canyon. Narrow canyons with steep walls are locations where a fire can jump across by radiant or convective spotting due to the short distances involved. These canyons are extremely dangerous for firefighters as the fire can easily cross over the canyon and get behind working fire crews.

Fuels
Fuel Loading
Size and Shape
Compactness
Horizontal Continuity
Vertical Continuity
Chemical Content

Figure 8-5 *Elevation changes fuel types.*

Saddles
The low topography between two high points.

Ridges
Elements that divide the terrain.

Chimneys
Steep, narrow draws in canyons.

Saddles are the low topography between two high points. They are the point of least resistance for wind and wind eddies. They provide the potential for rapid rates of spread because fires are pushed through saddles faster during upslope fire runs. Saddles are not to be used as a safety zone.

Ridges may not only divide the terrain, but also have totally different wind conditions on each side. This is especially true along coastal regions when weather patterns are changing due to the temperature difference between the cooler air from the ocean and the warmer air from the heated land areas.

Chimneys Steep, narrow draws in canyons are called **chimneys.** Radiant or convective heating can occur across narrow canyons due to the short distances involved. They draw fire up them (a drafting effect) in the same fashion as a flue draws heat up the chimney of a fireplace. A fire burning on level ground or up to a 5% slope will spread twice as fast when it reaches a 30% slope. The rate of fire

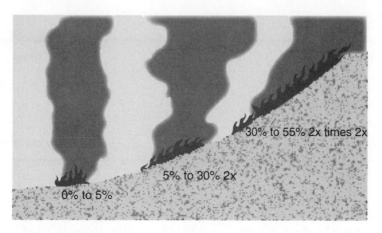

Slope/percentage

0% to 5%	x = spread of fire in feet per second
5% to 30%	2x = spread of fire in feet per second
30% to 55%	2x times 2x = spread of fire in feet per second

Figure 8-6 *Speed of increase as fire travels upslope.*

Fuel loading
A term that refers to the amount of fuel available to burn in a given area.

spread will double again as the slope reaches a 55% slope. Figure 8-6 depicts the speed at which you can expect fire to move upslope.

Fuels

Understanding and identifying the type of fuels is critical to firefighter operations and safety.

Fuel Loading The amount of fuel available to burn in a given area is termed **fuel loading.** When deciding on the tactics and strategy to use when fighting a structure fire, the firefighter calculates how much fuel is in the structure by the construction type and contents, termed the *fire load*. Based on this information, a determination of the amount of resources needed can be made.

The fuel loading of an area in a wild fire is determined by tons of fuel per acre, the fuel type, height and its density, the arrangement and shape of the fuel, and its moisture content and temperature. Table 8-2 illustrates how the fuel type relates to tons per acre. When a comparison of the fuel to the needed suppression activities is done, the fuel is generally classified as light, medium, or heavy.

Light Fuels. Grass and other small plants occur on the floor of all forests. These light flash fuels influence the rate of fire spread. If there is a continuous cover of dry grass on the forest floor, the rate of fire spread will be much faster. These fine,

Table 8-2   *Tons of fuel type per acre.*

Fuel Type	Average Estimated Tons per Acre
Grass	0.25 to 1
Medium brush	7 to 15
Heavy brush	20 to 50
Timber	100 to 600

Source: U.S. Forest Service, Fire Management Notes (2006).

light fuels ignite easily and provide an avenue for carrying the fire from one area to another and providing the heat necessary to move the fire to the heavier more dense fuels of the forest.

Medium Fuels. Medium fuels consist of **brush** that is 6 ft in height or lower, growing in fairly thick stands. In many cases, the medium fuels are those areas between the light and heavy fuels that act to move the fire from light to heavy fuels.

Heavy Fuels. Fires in the heavier (more mass) fuels produce more BTUs and as a result are more resistant to fire suppression activities. They require more water to cool, extinguish, and control them. They consist of brush taller than 6 ft, timber slash, and standing conifer and hardwood trees.

Fuel Shape and Arrangement   A fuel's shape may determine how the fuel can affect the ignition and spread of a fire. The smaller the fuel (the less mass), the faster it will preheat, ignite, and spread. The larger the fuel (the more mass), the more BTUs it will produce.

Fuels that grow and spread from the ground surface into higher vegetation are called ladder fuels. Ladder fuels support fire spreading from the ground area into the higher or taller vegetation.

Aerial Fuels   The **aerial fuels** include all green and dead materials located in the upper forest canopy. The main aerial fuel components are tree branches, crowns, snags, moss, and high brush. Aerial fuels can provide a path for fire travel in the higher fuels. These fires are inaccessible and as a result are difficult to extinguish. These fires travel through the top of fuels independent of the surface fire. They are called **crown fires.**

Fuel Moisture   The amount of moisture in the fuel will affect how easily it will ignite and how intensely it will burn. The moisture content of fuel is classified two ways. One is dead fuel moisture, which is the content of moisture in

Brush
Shrubs and stands of short scrubby tree species that do not reach large size.

Aerial fuels
Fuels that include all green and dead materials located in the upper forest canopy.

Crown fires
A fire that advances along tree tops and shrubs independently of the surface fire.

fuel not living. The second is live fuel moisture. The live fuel moisture is the amount of fuel moisture in living, growing fuels, which gain moisture primarily from the plant root system. Dead fuel moisture is changed only by the moisture content of the air. Both are measured as a percentage of the weight of the fuel.

The **time lag** is the time it takes for the moisture content of fuel and the surrounding air to equalize. Time lag is expressed as a rate usually in hours. Grass is considered a one-hour time lag fuel. It takes about one hour of air exposure to change the fuel moisture of grass up or down. This means that grass can go from 100% fuel moisture to a combustible fuel in one hour.

Logs are considered a 1,000-hour fuel. It takes about 1,000 hours of air exposure to an (higher/lower) air moisture content environment to change the dead fuel moisture in logs. Live fuel moisture is changed by the growing cycle of the vegetation and varies greatly between species and seasons.

Fuel Temperature The warmer the fuel, the less heat is required to ignite it. This is particularly important when dealing with light fuels. Their small size and mass allow them to heat up more quickly and change their moisture content faster (see Chapter 2 regarding mass).

There can be as much as a 50° to 80°F difference between the surface temperatures of fuel in the sun and those in the shade, with a corresponding difference in fuel moisture. Fire is a chemical reaction. For every 18° increase in temperature, the chemical reaction of fire doubles in speed. This influences both the intensity and the rate of fire spread.

Time lag
The time it takes the moisture content of fuel and the surrounding air to equalize.

Head
The outermost portion of the fire that is moving from the rear.

Anchor point
An advantageous location, usually a barrier to fire spread, from which to start constructing a fire line.

Island
An area inside the fire that has not been burned out.

Rear
The portion at the edge of a fire opposite the head.

FIRE BEHAVIOR

Identifying Parts of a Wildland Fire

Wildland fires are identified by a left flank (left side of the fire) and a right flank. The left side is always the left side of the fire when looking at a map. In other words, the left hand becomes the left flank of the fire. The **head** of the fire is the outermost portion of the fire that is moving from the rear (the bottom of the map, if the fire is not spreading south). The **anchor point** is a location near the origin of the fire from where it is safe to start working. Areas inside the fire that have not been burned out are called **islands.** If the area has not been burned on the perimeter it is called a pocket. It is surrounded on one side by a finger. The **rear** is the point of origin in most fires. Figure 8-7 identifies the critical parts of a wildland fire. Also note the spot fires and hot spots over and near the fire line. Firefighters always find a safe working area either using the anchor point or a safe location from which they start working on the fire line.

Figure 8-7 *Parts of a wildland fire.*

Convection column
The ascending column of gases, smoke, and debris produced by the combustion process of a fire.

Green area
The area next to a vegetation fire that has not been burned. Firefighters always watch for and protect against controlled fires moving into the green.

Spotting

The occurrence of spot fires depends on the topography and weather factors as well as the character of the fuels in the main fire. Spotting occurs when wind and **convection columns**—thermally produced column of gases, smoke, and debris produced by a fire—broadcast hot fire brands into the unburned fuel (**green area**) ahead of the main fire. Crown fires with their intense heat and strong convection currents frequently start spot fires; they can throw burning embers (preheating the brush) far ahead of the main fire front.

Large Fires

Very large fires can become so large that the combustion process creates its own indraft of oxygen, which increases and serves to continue the combustion process. Firefighters use this principle to backfire or burn out the fuel near the fire edge in an effort to separate the fire from the fuel. This fire extinguishment method works well when the backfire is ignited to be entrained into the main fire thus removing the fuel in the path of the main fire.

Area Ignition

Area ignition is an event that happens when spot fires occur in an area with unburned vegetation or when the spot fires are burning in a bowl or canyon or an area where the heat from the spot fires is trapped. In some cases, an inversion layer will hold the heated gases within the bowl. The spot fires that are burning preheat the unburned vegetation to a point where it reaches its ignition temperature. All at once, the preheated vegetation in the area explodes into combustion. If firefighters are in the bowl area, their chances of survival are slim. Similar conditions exist in a confined room where a flashover occurs resulting in the entire room exploding in fire.

Crown fires burn horizontally across the tops (crowns) of the trees or tall brush. These fires generally occur in fuels that are at least 6 ft (1.628 m) in height and when the fire is burning with a flame length anywhere between 30 and 200 ft (9.14 and 60.96 m). Crown fires generally move quickly. For a crown fire to occur, the vegetation must be close together (dense), continuous (one plant close to another), and be located where there is ground or ladder fuels to carry the fire into the tops of the trees.

If a ground fire has moved through an area without burning the crowns, a reburn in the form of a crown fire is still a possibility. Under this situation, the crown fire may actually be very intense, as the first fire through the fuel dried and scorched the aerial fuels.

FIRE RESOURCES

Engine companies, water tenders or trucks that carry water, hand crews, aircraft (both fixed wing and rotary wing), and bulldozers are used for wildland fires. Using the National Incident Management System (NIMS) in the command section, all of the resources are placed into a classification system based on their design, capability, and inventory of equipment carried. By using standardized classification systems, all agencies can quickly identify the type of equipment they are requesting from another agency or sending to another agency. This system standardizes the way resources are classified across the United States, allowing all agencies to understand the capability, design, and standard required equipment.

Field operations guide (FOG)
A small notebook designed to be carried in the shirt pocket, containing complete data on the ICS, including job titles, descriptions, and equipment classification systems.

The typing of all resources used on wildland fires is available in a **field operations guide (FOG).** The system classifies the units into strike teams and task forces. A strike team is a set number of resources of the same type with minimum staffing levels. Engine strike teams include five engines and a strike team leader, normally in a separate vehicle for mobility and communications purposes.

A task force is any grouping of resources with a task force leader, assembled on a temporary basis for a specific mission. A task force can be any size and include a variety of resources.

The engines within the strike team or task force are identified based on their capabilities by a classification system that puts them into types. The various engine types are not supposed to be mixed on strike teams; however, if they are in a strike team configuration, the team takes on the classification of the least capable unit.

Engine Types

Of the various types of apparatuses, both air and ground, the fire engine is the most versatile. It can fight fire both day and night, and is not restricted by visibility, as is the case with an aircraft. It can deliver water (attack the heat side of the triangle) to a fire at various capacities and transport a crew to a fire with their equipment. Its crew can construct hand line (attack the fuel side of the triangle) and the unit can be used to fight structural, vehicle, or other types of fire. Table 8-3 classifies the standard types of fire engines.

Hand Crew Types

On any sizable wildland fire, the biggest portion of suppression personnel will consist of hand crews. A hand crew is simply a group of firefighters formed into a suppression team. They use hand and power tools to reduce the fuel side of the triangle.

Hand crews usually contain from twelve to twenty firefighters who are assigned to actively suppress low flame production fires and construct hand lines along the fire's edge or ahead of the fire for future holding action or for backfiring operations. These crews often get the toughest assignments as they can get to locations where vehicles, including bulldozers, are unable to traverse.

Table 8-3 *Type classification of engines.*

Engine Companies	Pump (gpm)	Water Tank (gal)	2.5 Hose	1.5 Hose	1 Hose	Ladders	Staff
Type 1 engine	1,000	400	1,200	400	200	20 ft	4
Type II engine	500	400	1,000	500	300	20 ft	3
Type III engine	120	300	–	1,000	800	–	3
Type IV engine	70	750	–	300	300	–	3
Type V engine	50	500	–	300	300	–	3

A strike team of hand crews (more than one crew) consists of a strike team leader with two or more supervisors and up to forty crew members all of the same type. The kind of fire line work that can be performed by the crew is based on the type of the crew. There are two classifications of hand crews.

Type I: These persons are highly trained to work on all fires directly on the fire line.

Type II: These persons have some training with fire line work restrictions.

Hand crews can work with engine companies assisting in bringing a hose line or assist a bulldozer by following and picking up any fire missed by the dozer blade. Hand crews also need to perform cold trailing to make sure the fire is completely out.

Bulldozer Types

The use of bulldozers in conjunction with other ground components provides for a tactically balanced and integrated fire suppression force. Bulldozers attack the fuel side of the triangle. A strike team of bulldozers is defined as containing two dozers, two operators, two **swampers,** and a strike team leader. A swamper is a worker on a bulldozer crew who pulls the winch line, helps maintain equipment, and generally assists with suppression work on a fire. Bulldozers are typed into three broad classifications.

Type I: Heavy D-7, D-8, D-9—Heavy, thick brush, timber or road construction

Type II: Medium D-6 and HD-11—Heavy to medium brush

Type III: Light D-4 and HD-6—Light fuels, grass, fast-moving fire

Bulldozers are efficient and effective in removing fuels along a fire line. They do their best work during a direct fire attack by cutting a fire line on the fire flanks. On indirect attack they can be used to cut brush lines, escape routes, or safety zones in front of the main fire line. They also are used to open up roads and other access points in the wildland areas. Bulldozers are often used to cut water bars to ensure that soil is trapped rather than washed off the side of the hill during the rainy season. Figure 8-8 shows a fire fighting bulldozer.

Fixed-wing Aircraft Types

Both fixed-wing and rotary aircraft will attack the heat leg of the fire triangle, cooling with water or retardants. Aircrafts are typed on their ability to carry water or fire-retardant materials. Some are designed to drop the materials in a single drop or spread it out by dropping the load from their tanks in series, thus spreading the load over a larger area. Table 8-4 lists the classifications of fixed-wing aircraft and their carrying capacity.

Swamper

A worker on a bulldozer crew who pulls the winch line, helps maintain equipment, and generally speeds suppression work on a fire.

Figure 8-8   *Fire fighting bulldozer.*

Table 8-4   *Type classifications of fixed-wing aircraft.*

Type	Carrying Capacity (gal)
I	3,000+
II	1,800 to 2,999
III	600 to 1,799
IV	100 to 599

Fixed-wing aircraft can drop retardant or water on the head or front of a hot, fast-moving fire to slow it down and allow engine companies, hand crews, or bulldozers to work the flanks and eventually catch the head once it is cooled and moving slowly. In addition to slowing a fire, the aircraft can paint the fire line with fire retardant to slow the fire front or confine it within a specific area. Because they carry much more retardant than a rotary aircraft, the fixed-wing aircraft can cover a larger area which helps to reduce the total heat buildup and allows access for ground crews. Figure 8-9 illustrates a fixed-wing aircraft assisting in fire operations.

Rotary-wing Helicopters

The principle advantage of a helicopter is its ability to operate from locations close to and on the fire line. Its ability to take off vertically and land in close quarters makes it a valuable tool for close support action for ground-based and

Figure 8-9 *A Type III PBY5A fixed-wing aircraft which can drop water.* Photo Courtesy of FEMA/Sandy Bogucki/HHS.

Table 8-5 *Type classifications of rotary-wing aircraft.*

Type	Personnel Capacity	Agent Capacity (gal of retardant)
I	16 seats; 5,000 lb carded weight	700
II	9 to 15 seats; 2,500 to 4,999 lb	300 to 699
III	5 to 8 seats; 1,200 to 2,499 lb	100 to 299
IV	3 to 4 seats; 600 to 1,199 lb	75 to 99

other aerial operations. Various attachments that are available also permit a helicopter to perform certain jobs that would be difficult if not impossible by any other means.

Rotary-wing aircrafts cannot carry as much as the fixed-wing aircraft, but they can work closer to houses and pinpoint the application of the retardant more precisely. Nevertheless, both fixed- and rotary-wing aircrafts are needed as each unit fulfills an important but different fire protection function.

Helicopters may also be used for application of liquids, such as water or retardant chemicals, over key fire line targets. Because of their ability to hover, they can be directed precisely via radio communications with ground personnel. Table 8-5 classifies rotary-wing aircraft and their capacities.

In addition, helicopters can be useful in transporting injured firefighters and burn victims who will be flown to predetermined hospital burn wards. Helicopters have often served dramatically to evacuate persons stranded or threatened by fire. Figure 8-10 shows a rotary-wing aircraft.

Figure 8-10 *Rotary-wing aircraft.*

Direct Method of Attack

Direct method of attack
Attack that is directed to the edge of the burning fire.

The **direct method of attack** is where the attack itself is directed to the edge of the burning fire. Some ways to accomplish the direct method of attack are by water, chemical agents and retardants, or cooling—separating the main fire from the available fuel. For example, the water can be sprayed directly on the burning edge of the fire or a fire can be set to burn out the fuel in advance of the main fire. This method of attack has some tactical advantages and disadvantages for firefighters.

One tactical advantage is that the firefighters are working directly on the line and the actions result in the fire being controlled at the location where the attack is occurring. The line being cut around the fire must be continuous and complete enough to hold the fire to the burning side of the line. If successful, a direct attack limits the fire from building momentum and eliminates the need for backfiring or **firing out,** to "burn out" fuel between the main fire line and the line being constructed for fire control. It is generally used in the initial attack phase and used on the flanks or head of the fire depending on the amount of heat being generated. The main objective of this attack method is to stop the spread of the fire as soon as possible.

Firing out
Used during a direct fire attack to "burn out" fuel between the main fire line and the line being constructed for fire control.

Another advantage of the direct attack is that it holds the fire to a minimum acreage by taking advantage of line portions that have not burned due to lack of fuel or poor burning conditions. Other positive aspects include the fire being controlled more quickly with less loss of wildland resources than a backfire would require, and firefighters are able to use the black area to escape if necessary.

The disadvantages of the direct attack are that it may not be effective against an intensely hot or fast-moving fire and because this attack is risky, it requires close coordination of all crews. Generally, the incident commander makes this determination based on the flame length of the fuel, the topography, and weather conditions.

Because the fire line follows the burning fuel it is generally longer in length and is irregular as it follows the edge of the fire which causes a longer fire line construction. Another concern is that firefighters are working directly on the fire line so they are exposed to more heat and smoke while working on the fire line directly. They can accidentally spread burning materials across the fire line. Because the direct attack follows the fire line, it does not take advantage of barriers such as rock outcroppings, lakes, rivers, or other barriers.

Finally, there is usually more cleanup and patrol work needing to be accomplished and crews must carefully follow behind to ensure complete extinguishment.

Indirect Method of Attack

Indirect attack method
Consists of fighting a wildland fire by constructing control lines or by backfiring the fuel out at a distance ahead of the main fire.

The **indirect attack method** consists of fighting a wildland fire by constructing control lines or by backfiring the fuel out at a distance ahead of the main fire. It is used when the fire is burning too hot and too rapidly to employ direct attack methods or when insufficient suppression forces are not available. Because the indirect method is used on hot, intense fires, it is the choice most often selected on large fires.

A preconstructed fire line is put in ahead of the fire and is held in place by ground and air forces. If it can be held it becomes the final control line. Preconstructed lines can be reinforced with retardant drops to cool the fire as it approaches the line or it can be protected with a backfire.

A backfire is used to set fuel on fire in front of the main fire to deprive the main fire of fuel. A successful backfire stops the forward advance of the fire when the fuel in front of the fire is removed.

Advantages of the indirect attack method are the constructed control lines can be located using favorable topography. This line can often be completed with fewer suppression resources. In most situations, the total amount of fire line cut is shorter. Another important aspect is firefighters can work away from large amounts of heat and smoke and thus reduce their fatigue.

The disadvantages are that fire is not being extinguished during the time the control lines are being put into place and during this period additional fuel is being consumed. Also, any areas that are not burning are included in the control line perimeter so advantage is not taken of these lighter fuel areas and in some situations unburned fuel is left inside the control line. Further, it places firefighters in areas where burning out or backfiring is in progress. In some cases, secondary lines may need to be cut or burned out requiring more work and exposing firefighters to additional danger.

Combination attack
A method that takes advantage of the direct attack and the indirect methods when possible by combining both methods simultaneously.

Combination Attack

The **combination attack** takes advantage of the direct attack and the indirect methods when possible. Firefighters will work directly on the fire line in those areas that are safe and can be reached quickly in an effort to contain as much of the fire as possible. At the same time, resources permitting, crews begin to work indirectly in those areas where the fire line cannot be attacked directly to put in and burn out a clean line. Most incident commanders will use this method of attack as the opportunities on the fire line allow.

Application of Attack Methods

There is not one single method of fire attack that must be followed throughout fire suppression steps. Many times, a combination of two or more methods is used; or in some situations, all methods will be used simultaneously.

Black zone
The fuel or vegetation that has been burned. It is considered a safe area as the fuel has been removed.

Direct Attack (Tandem Action) This is a direct attack method with the attacking forces working in tandem (one unit following the other). The attack starts at an anchor point with units following closely on the flank of the fire moving up as the first unit knocks down the fire. This attack method is often used on a fast-moving fire as it is a direct attack on the fire line, and firefighters can always move from the fire line green area (the area not yet burned) into the **black zone** (the area already burned) to seek safety.

Pincer action
Moving crews along both flanks of the fire to a point where the flanking forces move closer together in a pinching action near the head of the fire.

Pincer Action A **pincer action** may be conducted on any size fire; however, it is most often used on small fires. The objective of a pincer action is to move crews along both flanks of the fire to a point where the flanking forces move closer together in a pinching action near the head of the fire. Resources include hand crews, bulldozers, engines, aircrafts, or a combination of these.

Flanking action
The fire is attacked on both flanks at the same time.

Flanking Action Any type of ground or air suppression resources can conduct a **flanking action.** The objective of a flanking action is to prevent the fire from spreading on a given flank and thus threatening exposures such as heavier fuel, a recreational area, or an area of structures. A flanking action requires sufficient resources to cover the entire fire area (both flanks) and good, close resource coordination. Flanking allows firefighters to work away from the direct flame front, reducing the amount of heat and smoke problems.

Envelopment action
Taking suppression action on a fire at many points and in many directions simultaneously.

Envelopment Action An **envelopment action** involves taking a suppression action on a fire at many points in many directions simultaneously. It provides for a rapid attack and, if coordinated properly, can be highly effective on smaller fires. This attack method requires a large amount of suppression forces to be committed to both flanks of the fire and close command and control must be

Parallel method
A method of suppression in which the fire line is constructed approximately parallel to and just far enough from the fire edge to enable personnel and equipment to work effectively.

established. Units taking part in this type of action should be experienced and aggressive.

Another method of attack is the **parallel method.** In this method of suppression, the fire line is constructed approximately parallel to and just far enough from the fire edge to enable personnel and equipment to work effectively, though the line may be shortened by cutting across unburned fingers. The intervening strip of unburned fuel is normally burned out as the control line proceeds but may be allowed to burn out unassisted where this occurs without undue delay or threat to the line.

SUMMARY

The original fire triangle model is applied rather than the newer fire tetrahedron model, as the fire triangle is used to explain both the combustion and extinguishment processes on wildland fires. Although wildland fires are similar to structure fires, because the same fire combustion principles apply, certain factors will influence differences in the fire behavior.

Structural fire behavior is impacted by the confinement of the combustion process within an enclosure. Wildland fires are outside fires where the weather, fuel, and topography affect the fire behavior. Each of these components affects fire behavior and under certain circumstances a combination of conditions in one or all can result in serious fire behavior.

The three methods of attack used on wildland fires are direct, indirect, and the combination attack. The type of attack depends on the fuel, weather, and topography as well as the type and availability of fire fighting resources. Each attack method has advantages and disadvantages. Other configurations of attack methods are pincer, flanking, and envelopment.

KEY TERMS

Aerial fuels   Fuels that include all green and dead materials located in the upper forest canopy.

Air stability and instability   The relationship between the vertical temperature distribution within an air mass and a vertically moving parcel of air.

Anchor point   An advantageous location, usually a barrier to fire spread, from which to start constructing a fire line. The anchor point is used to minimize the chance of being flanked by the fire while the line is being constructed. This may also refer to a safe ending point for a constructed line.

Area ignition   The ignition of an accumulation of heated gases from a number of individual fires in an area either simultaneously or in quick succession, and so spaced that they soon influence and support each other to produce a hot, fast-spreading fire throughout the area.

Aspect (exposure)   The direction a slope faces to the sun impacts the spread of fire in several ways. South facing slopes are exposed to more radiant heating by the sun, resulting in temperatures which are normally higher with lower humidity. Drier fuels are lighter and will ignite more quickly than the wetter fuels.

Black zone The fuel or vegetation that has been burned. It is considered a safe area as the fuel has been removed.

Bottom (toe) The rear-most area of a fire, generally used as an anchor or starting point for suppression activities.

Brush Shrubs and stands of short scrubby tree species that do not reach large size.

Burning index A number in an arithmetic scale determined from fuel moisture content, wind speed, and other selected factors that affect burning conditions and from which the ease of fire ignition and their probable behavior may be estimated.

Chimneys Steep, narrow draws in canyons. Radiant or convective heating or spotting can occur across narrow canyons due to the short distances involved.

Cold trailing A method of controlling a partly dead fire by carefully inspecting and feeling with the hand to detect any fire by digging out every live ember and trenching any live edge.

Combination attack A method that takes advantage of the direct attack and the indirect methods when possible by combining both methods simultaneously. A direct attack is used on the areas of the fire line that can be confronted safely, while areas that are considered too dangerous for this method are attacked using indirect methods.

Convection column The ascending column of gases, smoke, and debris produced by the combustion process or a fire.

Crown fires A fire that advances along tree tops and shrubs independently of the surface fire. Sometimes they are classed as either running or dependent, to distinguish the degree of independence from the surface fire.

Direct method of attack Attack that is directed to the edge of the burning fire.

Envelopment action Taking suppression action on a fire at many points and in many directions simultaneously. This action provides for a rapid attack and, if coordinated properly, can be highly effective on small fires.

Evaporation The process by which molecules in a liquid state (e.g., water) spontaneously become gaseous (e.g., water vapor). Water vapor is simply water in a gaseous state and is an important ingredient in fire behavior.

Field operations guide (FOG) A small notebook designed to be carried in the shirt pocket, containing complete data on the ICS, including job titles, descriptions, and equipment classification systems.

Fire retardant Any substance that by chemical or physical action reduces flammability of combustibles.

Firing out Used during a direct fire attack to "burn out" fuel between the main fire line and the line being constructed for fire control.

Flanking action The fire is attacked on both flanks at the same time. The objective of a flanking action is to prevent the fire from spreading on a given flank and thus threatening some exposure such as heavier fuel, a recreational area, or an area of structures.

Flash fuels Fuels such as grass, leaves, ferns, tree moss, and some types of slash which ignite readily and are consumed rapidly when dry.

Foehn A dry wind with a strong downward component locally called Santa Ana, North, Mono, Chinook, or East Wind.

Fuel loading A term that refers to the amount of fuel available to burn in a given area.

Fuel type An identifiable association between fuel elements of distinctive species, form, size, arrangement, or other characteristics that will cause a predictable rate of fire spread or difficulty of control under specified weather conditions.

Gravity winds Winds that occur when air is pushed over high elevations and flows downhill. They are identified in a number of ways depending on their location, including foehn winds, devil winds, sundowners, and Chinook winds.

Green area   The area next to a vegetation fire that has not been burned. Firefighters always watch for and protect against controlled fires moving into the green.

Head   The outermost portion of the fire that is moving from the rear.

Indirect attack method   Consists of fighting a wildland fire by constructing control lines or by backfiring the fuel out at a distance ahead of the main fire.

Interface lands   The areas where structures interface with wildland vegetation.

Inversion layer   A layer of comparatively warm air overlaying cool air. In this stable atmospheric condition, the atmosphere will resist vertical motion of the air.

Islands   An area inside the fire that has not been burned out.

Parallel method   A method of suppression in which the fire line is constructed approximately parallel to and just far enough from the fire edge to enable personnel and equipment to work effectively.

Pincer action   Moving crews along both flanks of the fire to a point where the flanking forces move closer together in a pinching action near the head of the fire.

Rear   The portion at the edge of a fire opposite the head.

Red flag alert   The condition that occurs when relative humidity is below 15%, causing fuel to become very dry and thus making conditions favorable for a wildland fire. Fire departments then prepare for wild fires by moving and adding additional resources and by activating special fire prevention activities.

Ridges   Elements that divide the terrain. Different wind conditions may be present on each side of the ridge. This is especially true along coastal regions when weather patterns are changing due to the temperature difference between the cooler air from the ocean and the warmer air from the heated land areas.

Saddles   The low topography between two high points.

Spot fires   Fires that jump over a control line and start the vegetation burning ahead of the main fire front.

Swamper   A worker on a bulldozer crew who pulls the winch line, helps maintain equipment, and generally speeds suppression work on a fire.

Temperature inversion   An action that occurs when cooler air is found close to the ground surface and warmer air is above this pool of cooler air.

Time lag   The time it takes the moisture content of fuel and the surrounding air to equalize.

REVIEW QUESTIONS

1. Do you believe the basic fire and combustion processes found in wildland fire fighting differ from structural fire fighting? Explain your answer.

2. Explain the similarities and differences between the strategy and tactics of wildland fire fighting and structural fire fighting.

3. While working to extinguish a wildland fire, how would you exclude the oxygen, reduce the heat, and remove the fuel?

4. List the three main factors that impact the behavior of a wildland fire and include how each impacts fire behavior.

5. Why are chimneys so dangerous to firefighters?

REFERENCES

Barr, Robert C. and John M. Eversole. (eds.). (2003). *The Fire Chief's Handbook* (6th ed.) (Saddle Brook, NJ: Penwell Publishing Company).

Boise Interagency (fire information and publications), *http://www.nifc.gov/fire_info/nfn.htm.*

Clayton, B., D. Day and J. McFadden. (1987). *Wildland Firefighting* (North Highland, CA: State of California, Office of Procurement).

"Effectiveness Fire Pooled Fire Resources on Wildland Fires." (1989). Report to the Los Angeles County Board of Supervisors.

Klinoff, Robert. (2007). *Introduction to Fire Protection* (3d ed.) (Clifton Park, NY: Delmar Cengage Learning).

Lowe, Joseph D. (2001). *Wildland Firefighting Practices* (Albany, NY: Delmar Cengage Learning).

Perry, Donald, G. (1990). *Wildland Firefighting: Fire Behavior, Tactics & Command* (2d ed.) (Bellflower, CA: Fire Publications).

Queen, Phillip L. (1993). *Fighting Fire in the Wildland/Urban Interface* (Bellflower, CA: Fire Publications).

Teie, William C. (1994). *Firefighter's Handbook on Wildland Firefighting: Strategy, Tactics and Safety* (Rescue, CA: Deer Valley Press).

U.S. Department of Agriculture Forest Service. (1994, rev. 2001). Federal Wildland Fire Policy, *http://www.fs.fed.us/fire.*

U.S. Fire Administration and FEMA. (1991). *The East Bay Hills Fire Oakland-Berkeley, California (October 19-22).* Technical Report No. 060 *https://www.usfa.dhs.gov/applications/publications/display.cfm.*

U.S. Forest Service. (2006). *Fire Management Notes* (Boise, ID: U.S. Government Printing Office).

Chapter 9

TRANSPORTATION FIRES
AND RELATED SAFETY ISSUES

Learning Objectives

Upon completion of this chapter, you should be able to:

- Examine fire behavior and safety-related problems in transportation vehicles encountered by firefighters.
- Describe fire problems and safety issues experienced with transportation vehicles and explain actions that may be taken to resolve the issues.
- Examine and describe special fire behavior problems one might encounter with each of the classifications of transportation vehicles.
- Explain the importance of fire preplanning and familiarization procedures for each of the categories of transportation vehicles.

INTRODUCTION

■ Note

Responders should consider the presence of hazardous materials for every transportation emergency as the hazardous materials may not only be found as the cargo but can also be found in the materials used to manufacture the vehicle such as composites, plastics, and lubricants.

Fires and related emergencies occurring in transportation vehicles are examined in this chapter. Concerned with the growing number of transportation vehicle deaths and injuries, the U.S. Fire Administration in 2003 funded an examination by the National Fire Protection Research Foundation into transportation-related deaths and injuries to identify the number and impact of these incidents. This report is used as the basis for understanding the depth of the nation's transportation vehicle fire and related safety issues.

The report found that during the years from 1997 to 2002, an average of 399,939 fires occurred in transportation vehicles, as shown in Table 9-1. Each year these fire incidents resulted in the average of 587 fire deaths, 2,347 fire-related injuries, and $1.24 billion in fire losses.

Because of the wide variety of transportation vehicles under the vehicle classification system used by many organizations, the number of vehicle categories has been reduced to cover only the major vehicle classifications. The categories include passenger car fires, truck and recreational vehicle fires, rail transportation vehicle fires, marine vehicle fires, and aircraft fires.

Specific fire problems are identified as well as any unusual fire behavior and safety issues facing first responders. Although fire principles and suppression tactics are similar to structural and wildland fires, certain unusual and dangerous fire behavior situations arise from transportation vehicles fires.

Table 9-1 *Transportation vehicle deaths, injuries, and losses.*

Type	Fire	Percentage	Fire Deaths	Percentage	Fire Injuries	Percentage	Fire Loss (Billion $)	Percentage
Passenger	292,170	73.8%	330	56.2%	1,403	59.8%	$692	55.8%
Freight road	37,100	9.3%	104	17.7%	345	14.7%	$183	14.7%
Rail transportation	650	0.02%	6	1.0%	12	0.5%	$21	1.7%
Water transportation	1,540	0.4%	6	1.0%	73	3.1%	$20.9	1.7%
Air transportation	200	0.1%	38	6.5%	25	1.1%	$39.7	3.2%
Heavy equipment	6,260	1.6%	7	1.2%	65	2.8%	$77.2	6.2%
Special vehicle	2,200	0.6%	2	0.3%	29	1.2%	$9.3	0.7%
Unknown	56,810	14.2%	94	16.0%	395	16.8%	$197.2	15.9%

Source: USFA, 2004.

Note: After reviewing the results of this study, many fire officials have become concerned. Data confirm the fire problem of transportation vehicles here in the United States, which is much more significant than most U.S. citizens want to believe it is. When we compare these numbers with the numbers of structural fires during the same period, we find the number of vehicular fires is almost two-thirds of the number of structure fires (USFA, 2004, p. 3).

PASSENGER VEHICLE FIRES

Passenger vehicles differ from other classifications as these vehicles can be brought to a stop rapidly and evacuated almost immediately. Using this basic premise, the **National Highway Traffic Safety Administration (NHTSA)** developed a small-scale regulatory fire test for vehicle interior materials in the late 1960s and has continued applying the same test today.

During the 1960s and 1970s, many of the materials used in private passenger vehicles were noncombustible. Recently, there has been considerable growth in the use of combustible plastics, increasing the fire danger without a concern for fire safety.

The data from the 2000 USFA transportation vehicle report shows annually passenger road vehicles were responsible for almost 60% of fire-related fatalities occurring in transportation vehicles. This study reports that fires and safety issues in newer passenger vehicles is a larger problem than most of us believe. While a great deal of research has been devoted to improving the safety record of these vehicles, some of the safety devices have created safety and fire problems for firefighters.

Vehicles built prior to the 1960s were routinely extinguished with small booster lines that could be applied quickly as they provided sufficient water for most fires. Occasionally a fuel tank was involved in fire or the tank was damaged and flammable liquid escaped creating a problem requiring more water than a booster line could supply.

In addition to burning hotter and producing toxic smoke, these newer cars also have many more systems operating under pressure. If heated sufficiently, they will explode during a fire. One example is the new air pollution control regulations which require the fuel system to be pressurized to control the release of vapors. Some of these systems use plastic fuel supply lines which are pressurized from 15 to 90 psi.

The fuel tank is made of polyethylene or polypropylene plastic. These tanks are lighter and more durable than steel tanks, but will melt after a few minutes of flame contact. Because they also are pressurized, a rupture can result in spraying flammable liquid on firefighters working suppression lines. In 1990, one firefighter was seriously burned when fighting a gasoline-fed fire near the rear compartment of the passenger vehicle. The firefighter working near the fuel tank did not realize that the plastic fuel cap was pressurized and the fuel sprayed out of the fill pipe, burning his face.

Front Bumpers

The hydraulic cylinders on front and rear bumpers are designed to absorb a 5 mph crash test without damage; however, firefighters need to take caution as bumpers can explode or react during the operation of an extrication tool under the intense heat given off by a flammable liquid fire.

National Highway Traffic Safety Administration (NHTSA)
A group that developed a small-scale regulatory fire test for vehicle interior materials in the late 1960s and has continued applying the same test today.

■ **Note**
An average vehicle fire requires a minimum 1.5-inch (37-mm) line or 1.75-inch (45-mm) line to supply the amount of water because of the significant increase in the use of plastics. Plastic- or hydrocarbon-related fires burn twice as hot as wood-based fires and produce a very dense, toxic smoke which further hampers the fire fighting activities. As a result, firefighters must absolutely wear self-contained breathing apparatus (SCBA) and full protective clothing to safely suppress vehicle fires.

Air Bags

Supplemental restraint system (SRS)
An additional passenger restraint system, some of which are placed in the vehicle roof, dash, or in a roof pillar.

Side impact curtain (SIC)
A new vehicle passenger restraint system that deploys more quickly and forcefully from the side because of the shorter distance to the passenger.

Inflator
A pressurized container that holds gas that is released upon impact, filling the curtain or bag which restrains passengers within the seating area.

Another problem facing first responders is the accidental deployment of active airbags during vehicle rescue. Newer vehicles have a **supplemental restraint system (SRS)** which must be considered when cutting through undeployed inflators hidden in a roof pillar, dash, or behind the roof liner trim of the vehicle interior occupant compartment.

The **side impact curtain (SIC)** is one of the latest innovations offered by the auto industry for side impact and rollover protection. Side impact protection systems (SIPS) react more quickly and deploy more forcefully than frontal restraint systems because of the shorter distance between the point of impact and the occupant. The **inflator** produces large volumes of cool/hot pressurized inert gases that fill the appropriate impact curtain (bag) during the deployment stage. The pressure vessel in hybrid inflators will have a stored gas static pressure ranging from 3,000 psi to more than 4,000 psi. It is possible that firefighters could inadvertently breach (rupture) a hidden impact inflator while displacing or removing a roof.

At fires along roadways, firefighters often are in more danger of being struck by another vehicle than of being injured by the fire. To provide some protection, fire departments often have policies and procedures in place to establish safe work areas while on the roads or highways. These often involve positioning of the apparatus, using warning and marking devices to alter traffic flow, and using personnel to assist with monitoring and directing traffic flow. Figure 9-1 depicts an example of how a safe work area is created at the incident scene. The fire apparatus creates a traffic barrier while cones and a spotter/flagger slow traffic. Figure 9-2 is an example of a safe work zone created at an emergency incident.

Firefighters attack a vehicle fire targeting the passenger compartment first in order to protect persons trapped inside the vehicle.

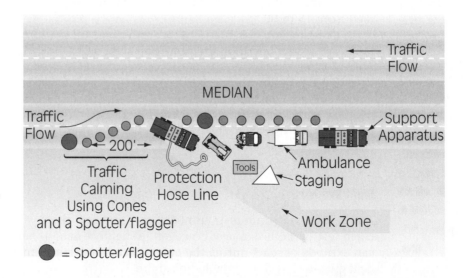

Figure 9-1 *An example of how a safe work zone is created on the incident scene.*

Figure 9-2
Firefighters in full protective gear and SCBA attack a passenger vehicle fire with hand lines.

■ **Note**

Cutting into the electrical wire harness could provide an open avenue for an electrostatic charge to ignite the IC inflator. Therefore, firefighters should avoid cutting into any wiring harness when possible, as damage can allow an outside static electrical charge to ignite the inflator.

Hybrid Passenger Vehicles

Firefighters need to be alert as some of the newer **hybrid passenger vehicles** (Figure 9-3) may contain hydrogen fueling systems, compressed natural gas (CNG), or compressed liquefied petroleum gas (LPG) in pressurized tanks. Other vehicles use electrical powering systems with voltage as high as 600 volts, and some use lithium batteries.

A recent vehicle fire in North Carolina challenged firefighters as they faced ten separate batteries charged to a total of 336 volts with 100 amps of electrical current. The burning lithium batteries added to the firefighters' concern. Firefighters need to become familiar with these new power and fueling systems as they dramatically impact the behavior of fire and safety problems presented by these vehicles.

Figure 9-3 depicts a hybrid vehicle. Due to the recent demand issues with gasoline and diesel fuel, some vehicles use hydrogen gas, others use a combination of natural gas and liquefied petroleum gas, and others use a combination of gases with electric motors/batteries. One vehicle uses refined cooking oil, resulting in a greasy and fire-prone vehicle. Firefighters of the future will need to be experts in all types of vehicle fueling systems.

Hybrid passenger vehicles

A term used to describe vehicles powered by a variety of fuels and powering systems.

Problems Presented by Alternative Fuel Vehicles When responding to a vehicle fire, firefighters must be aware that it may be powered by alternative fuels. The fuels are **compressed natural gas (CNG)** or liquefied propane gas (LPG). Fires in these vehicles may require a fire attack at a distance as the pressurized tanks holding these gases may explode.

Automobiles powered by CNG contain storage cylinders that are similar to SCBA cylinders. Located primarily in the trunk or cargo area, these cylinders

Figure 9-3 *Hybrid vehicle.*

Compressed natural gas (CNG)
Natural gas that is compressed and contained within a pressurized container.

contain CNG at high pressures. They must be cooled and protected as with any gas-filled cylinder. CNG is a nontoxic, lighter-than-air gas that will rise and dissipate if it is released into the atmosphere.

Propane is stored in the same type of cylinder as those used for heating and cooking purposes. Propane is heavier than air, so the vapors will pool or collect in low areas. Table 9-2 provides a reason for research work to find a suitable

Table 9-2 *Heat of combustion of commonly used fuels.*

Fuel	BTUs/lb	kJ/kg
Hydrogen	61,600	141,800
Methane (natural gas)	24,100	55,500
Propane	21,500	50,000
Diesel fuel	20,700	48,000
Gasoline	20,000	46,800
Cooking oils	14,000	34,700
Methanol	9,900	22,680
Wood	8,500	19,700

Source: Adapted from National Fire Protection Association. (2008) Fire Protection Handbook (20th ed). (National Fire Protection Association), pp. 6–270.

■ **Note**
Although the standard for automotive electrical systems is 12 volts, some have operating systems with a voltage up to 600 volts. This boost in voltage increases the possibility for ignition of fuel that has been released during an accident. If flammable fuels are involved, batteries need to be disconnected while other ignition sources are prohibited within the immediate area.

alternative fuel source. The amount of energy that can be released from hydrogen is three times that of gasoline without the environmental impact as hydrogen burns free of toxic gases. Natural gas (CNG) and propane/butane (LPG) produce about an equal amount of energy output of gasoline without the toxic combustion products.

Hydrogen is a basic chemical element that is very light, weighing only one-fifteenth as much as air with an extremely wide flammable range (4% to 75%). It also has a high heat content per pound (see Table 9-2). It is colorless, odorless, and contains no carbon products, causing it to burn with a nonluminous flame which produces little or no toxic by-products. This is one reason why it was selected to be researched as a fuel alternative to petroleum-based products. When released from containment, hydrogen presents both an explosion and a fire hazard with its wide flammable range, which is the highest of all of the flammable gases. Figure 9-4 shows a picture of a hydrogen-powered car.

Vehicle Electrical Systems In a number of emergency situations involving vehicles, passengers will become trapped and require the fire department to provide extrication services. Care needs to be taken to be sure the battery is disconnected while these operations are being conducted so an ignition of leaking flammable liquids does not occur.

The experimental electric vehicle in Figure 9-5 operates at 600 volts and high amperage. The high voltage and high amperage cause such cars to be fire prone. This type of vehicle has a small gasoline engine with a large bank of batteries. The batteries power electric motors that drive the wheels, similar to a train locomotive. The batteries are nickel metal hydride and are hazardous when on fire as they sometimes explode. Water used to fight these battery fires becomes contaminated; it becomes a hazardous material and must be contained and disposed of as required by law.

Cables are used to connect the batteries to the electric motors powering the wheels. Cutting these cables may result in serious injury or death as the voltage is as high as 600 volts. The cables are installed from the battery bank in the rear to the front of the car via the undercarriage. These cables pass directly under the center of the driver's seat. Extra care should be taken when using hydraulic metal cutters or spreaders (rescue tools) on these types of vehicles to avoid cutting the high-voltage cables.

Strategy and Tactics in Passenger Vehicles

Always respond with sufficient resources to ensure ready availability in case the vehicle contains materials and/or trapped persons.

The first step is to carefully examine the situation using the decision-making process to gather the needed information for developing a plan for safe strategy and tactics. If the fire can be safely attacked and sufficient water is available, the attack should be started from the downwind side and upslope, if possible.

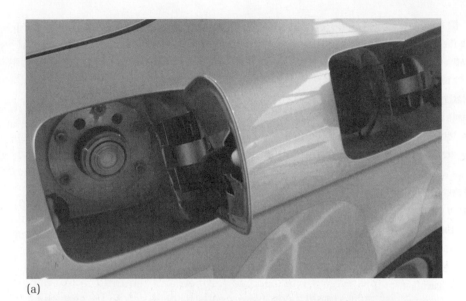

(a)

(b)

Figure 9-4
Hydrogen-powered car. Photo
(a) Courtesy of
Claus Ableiter.

Figure 9-5 *New hybrid vehicle (electric).*

Firefighters should use caution to prevent leaking flammable fuels from spreading into other areas of the vehicle as persons may be trapped. Leaking fuel can spread into low areas or storm drains surrounding the vehicle as well. In some cases, foam can be effective in this situation to provide extinguishment and containment and act as a flammable vapor blanket.

MOTOR HOMES, BUSES, AND RECREATIONAL VEHICLES

A recreational vehicle contains living space for travel or camping on a seasonal basis. The variety of vehicles included in the classification of recreational vehicles ranges from a small pop-up tent trailer to a large buslike vehicle containing all the amenities of home. These resources for cooking, bathing, and lighting pose the same hazards found in residential housing units. The only difference is recreational vehicles, as well as motor homes, are mobile and the fuel lines and electrical connections are subject to vehicle accidents and vibration as the vehicle moves along the highway.

Mobile homes are classified into three categories: Class A, B, or C. The Class A units are designed and built as mobile homes. Class B units have been converted from a basic van bought from a vehicle manufacturer. Class C motor homes are units that use the cab portion and motor from the vehicle manufacturer with the addition of a body generally containing an over-the-cab sleeping unit.

In terms of fire risks, these vehicles can be considered a residential property on wheels, containing the same combination of risks one will find in residential properties. Many are equipped with heaters, stoves, air conditioners, and appliances. To power the appliances, the vehicles have a combination of electrical batteries, ranging from 12 to 24 volts direct current (DC), with switching mechanisms to utilize a 115-volt alternating current (AC) system. The AC system can be fed by an onboard electric generator or a land-based electrical supply cord. These systems should have circuit breakers installed for protection against electrical overload.

Heating and cooking energy can be supplied to the appliances by onboard LPG tanks ranging from 5 to 80 gallons. All LPG tanks are equipped with some type of pressure relief valve designed to vent the contents of the tank at a pre-designated pressure. When an LPG tank heats due to flame impingement, the pressure inside the tank increases. At the point when this pressure is greater than the relief valve setting, the gas is vented. This venting continues until the pressure inside the tank is less than the pressure relief setting. Once the pressure is reduced, the relief valve closes. If the fire continues to heat the tank, the valve will remain open and continue to expel all of the tank's contents. These flammable vapors are heavier than air and will seek the lowest point awaiting an ignition source. Recently, a family of six was killed when their mobile home exploded from a leaking LPG tank improperly connected to a cooking stove that was installed by the father.

Motor home interiors are furnished with the same type of furniture, carpeting, and appliances as in a house. These furnishings contain a great deal of plastics, synthetic fabrics, foam rubber, and particleboard to save both cost and weight; however, their installation significantly increases the amount of fire load per square foot.

The structural components of mobile homes consist of lightweight aluminum tubing for the framing with a lightweight aluminum skin stretched over the framing. Plywood or particleboard is used inside the walls and floors with polystyrene or polyurethane foam sandwiched between layers of the plywood or particleboard. This combination offers little fire resistance and once ignited there is little to stop the entire structure from being consumed and collapsing. The products of combustion are toxic as well, producing formaldehydes and a variety of cyanogens, all of which can be fatal.

Frequently, passengers in motor homes do not use seat belts and when a vehicle accident occurs, these individuals are thrown about inside the mobile home creating a need for medical services. Firefighters may be surprised to find entire families injured. In one recent incident, eleven people inside an overturned motor home required medical services, and three people were killed. Figure 9-6 shows a Class A motor home.

Figure 9-6   *Class A motor home.*

Fire Tactics and Strategy in Motor Homes, Buses, and Recreational Vehicles

Fires in buses are not much different from fires encountered in large recreational vans or trucks. Like all other vehicles examined, there is a wide variety of configurations designed to transport people. For example, there are school buses, public transportation buses, and group charter buses as well as commercial bus lines. Because of the possibility of persons being trapped inside the vehicle by a fuel-fed fire or vehicle accident, quick fire attack and rescue extrication may be needed.

While most buses are powered by diesel engines, many publicly operated buses are now being powered with natural gas or methane. Methane is the simplest hydrocarbon. Its chemical formula is CH_4. Methane is odorless, colorless, and tasteless and is the primary component of natural gas. It is a desirable fuel when compared to gasoline not only because it provides a high heat value, but also because its combustion produces less volatile organic compounds compared to the combustion of gasoline. For these reasons, many cities have elected to use compressed natural gas as a fuel in the city-operated buses. (See, for example, Figure 9-7, which shows the compressed gas cylinders on top of the bus.)

Methane is lighter than air so the fuel storage location on top of the bus is a relatively safe location. Any leakage will rise into the atmosphere and firefighters do not need to concern themselves with a leaking fuel that may trap passengers.

Figure 9-7 *Bus fueled by compressed natural gas.*

(a)

(b)

TRUCKS

Trucks are vehicles that are designed to carry some type of cargo. The cargo varies greatly in size and shape as well as hazard. Cargo can be explosive or toxic and may not always be identified with a placard or warning label. Trucks are designed to carry goods great distances and as a result many have a sleeping compartment, which could be exposed to fire from inside the compartment (smoking) or from the cab or cargo area.

The majority of international commodities transportation is shipped in containers, which can easily be unloaded from a ship and transferred quickly to a truck for ground transportation. These containers carry a wide variety of cargo, some of which may be hazardous and not marked as required. See Chapter 10 for information on international placard requirements.

Challenges Encountered in Truck Fires

A number of potential fire and safety hazards may be encountered with the variety of goods carried by trucks. Firefighters must always proceed with fire extinguishing operations using caution until the cargo can be positively identified.

Even if a nonhazardous cargo is burning in a large trailer, accessing the trailer can be dangerous as many trailers have only two double doors on the rear with no side access. This means firefighters are limited in the ways they can approach the fire during extinguishment of the interior cargo space (see Figure 9-8).

Figure 9-8   *Cargo box on a tractor trailer.*

Figure 9-9   *Exposed fuel tank.*

Saddle tanks are typically found on large trucks (Figure 9-9). These are large fuel tanks located on the side of the cab. They are usually under the driver and passenger doors. They range from 50 to 250 gallons (227.3 to 1,136.5 liters) each. The saddle tanks are vulnerable as they are mounted to the outside of the

Cribbing
Blocks of wood or prefabricated plastic blocks used to lift a vehicle's frame off the wheels and stabilize the vehicle.

body frame and are exposed to damage from vehicle accidents and road debris. Leaking fuel tanks can present a serious fire problem, as flowing fuel fires must be dealt with quickly because they can rapidly spread fire beyond the immediate area. They are a serious life hazard to firefighters working in or around flowing fuel. The fuel tank is attached directly to the truck frame and is exposed to a side collision. The air supply tank for the braking system is located behind the fuel tank. Collision from the side could split the fuel tank and open the air brake system, which would automatically activate the braking system.

Newer, large tractor trailers may have molded fiberglass air skirts to reduce air resistance as the truck moves down the highway. They can obscure the location of the saddle tanks and their close relationship to the battery. Prior knowledge of trucks and their design is helpful to firefighters. A preplanning walkthrough at the local truck dealership can be helpful in learning to locate and understand vehicle fuel and electrical systems.

Some trucks are equipped with air suspension systems; failure of this system releases the air pressure causing the trailer to drop several inches. Firefighters working to extricate a victim under a trailer with air suspension should set **cribbing** in place before any rescue work is started. Figure 9-10 depicts cribbing being used to hold a vehicle frame off the wheels and stabilize the vehicle once the frame is lifted with a hydraulic jack. The blocks can be made of wood or strong prefabricated plastic materials.

Truck Brake Fires

Firefighters will occasionally encounter a truck with its brakes on fire. The brakes can become overheated when applied. This is of special concern when the vehicle is traveling in mountainous areas and the heavy loaded vehicle is descending a long, steep declining grade. When heated, the metal brake drums

Figure 9-10 *Cribbing used to hold and stabilize the lifted vehicle.*

can crack if cooled too quickly. Firefighters should apply fog spray in short bursts to slowly cool the heated drums.

Fire Tactics and Strategy in Trucks

Bill of lading
Documents indicating the amount and type of cargo or freight being transported.

In most cases, the cargo area will be of concern on a truck or trailer fire or accident. Always check for hazardous materials placards and do not take for granted that the warning placards will be posted or that the truck has proper placards. Also, trucks will have a **bill of lading** which will indicate the amount and type of cargo or freight being transported.

A few years ago, firefighters responding to a fire at a construction site were told that explosives may be on-site but the information was not specific in regard to location. Firefighters started suppression activities not knowing that the burning truck trailer (without warning placards) contained explosives (USFA, 1988). Firefighters were unaware that the truck trailer/magazine contained a mixture of ammonium nitrate and fuel oil with aluminum pellets. The truck trailer containing the mixture was exposed to fire resulting in an explosion that killed six firefighters.

FIRES IN RAILED EQUIPMENT

Although a fire department may have outstanding knowledge of its local building target hazards and have completed excellent prefire plans for dealing with these situations, some of the most dangerous problems a department may encounter are the freight trains which pass through the jurisdiction daily. Any fire department that covers an area where train tracks pass through has the potential to respond to a rail incident. Trains, like ships, can carry large amounts of people or large amounts of cargo and in some cases a cargo may be hazardous. Light rail and subway systems, even though they do not typically carry cargo, can have a high life hazard when involved in fire because of the large numbers of people they carry every day.

Railcar Construction and Placards

The rail transportation of hazardous commodities is regulated by the U.S. Department of Transportation (DOT). These regulations are enforced by the Federal Railroad Administration (FRA) and can be found in *Hazardous Materials Regulations* 49 CFR part 179. In addition, the design and construction of the containers to transport hazardous materials is carefully regulated in specifications set forth by the Association of American Railroads (AAR, 1987).

Rail transportation is a specialized business with special procedures, regulations, and equipment. It is highly recommended that firefighters use preplan guidelines to visit and preplan any rail facilities they may have in their jurisdiction. For those fire departments that do not have rail facilities in their jurisdiction but have rail traffic running through their jurisdictional area, specialized drills and seminars can be arranged with rail officials.

Locomotives

Locomotives are the power producers for the train. The diesel electric locomotive is a diesel-powered generator that creates electricity to power the train. To fuel the diesel engines, cross country trains carry up to 5,000 gallons (18.92 kiloliters) of fuel. A ruptured diesel tank can create a serious fire problem.

There is a small chance of fire in the diesel engine itself. Most of the hazard is the electricity being generated by the large electrical generators driven by the diesel engines. Newer locomotives control the electrical power for the train with special designed switches and automated circuit breakers.

Another factor making trains dangerous is their large size and weight. These provide inertia energy, so once the train is set in motion, a long distance is required before it can be stopped. Trains have a **consist** or **way bill** which identifies hazardous cargo onboard. The engineer of the vehicle has the responsibility to maintain these documents. The documents can be found generally in the cab of the vehicle.

Consist (way) bill
Documents used to identify the hazardous cargo of a rail transportation vehicle.

Boxcars

Boxcars get their name from their appearance, as they look like a box with attached wheels. They are designed to carry a variety of commodities which may or may not be hazardous and will have varied levels of combustibility. Boxcars are designed to protect the cargo from weather elements and contamination. Often they are made of wood, which adds fuel to fire once the car is ignited. Some boxcars may contain a refrigeration system. The compressor is electric and is powered by a diesel generator attached to the boxcar; the fuel for the generator is carried in a small fuel tank attached to the car.

Boxcar
A railroad car that is enclosed and used to carry general freight.

Flatcars

Flatcars are not enclosed so they do not provide protection from the weather. They are designed and configured in different sizes depending upon the type of cargo they are designed to carry. Some flatcars are constructed from wood materials so the main fire concern is the wooden car and the cargo.

Flatcar
A railcar that is open to the elements.

Intermodal Equipment

Intermodal railcars are flatcars with an intermodal container attached. The design of the container allows it to be transported and transferred quickly between

Intermodal railcar
Transportation of freight in a container or vehicle, using multiple modes of transportation (rail, ship, and truck), without any handling of the freight itself when changing modes.

Figure 9-11 *An IMO 101 tank. Like the totes, these are bulk tanks capable of carrying a large quantity of product. They are normally placed on ships and then delivered locally by a truck, although trains can also be used.*

a ship, truck, or rail carrier. For example, an intermodal cryogenic container could be shipped by sea to a party, transferred by crane to a truck, trucked to a rail yard, loaded onto a flatcar, transported to a rail yard across the nation, and loaded back onto a truck for a local delivery without ever having to off-load the product.

The hazards associated with intermodal equipment are related to the cargo itself. The cargo varies with a wide range of hazardous materials such as flammable gases, greases and liquids, or pressurized gases and cryogenics. See the warning placards and labeling requirements in Figure 9-11. Note that the letters "UN" or "NA" may precede the identification numbers. According to the *Emergency Response Guidebook* (DOT, 2004) regulations, "those preceded by the letters UN are associated with proper shipping names and are considered appropriate for international transportation and domestic transportation, those proceeded by the letters NA are associated with proper shipping names not recognized for international transportation except to and from Canada." An intermodal tank is commonly referred to as an IM or IMO.

Gondola Cars

Gondola cars
Railcars designed with flat bottoms and four walls that can carry a variety of goods.

Gondola cars are designed with flat bottoms and four walls. They may have a cover for the cargo being carried. Some are constructed of wood, but most often are constructed of steel. Those constructed of wood can be a fire problem with some commodities but in most cases the only fire concern is the hazard of the cargo itself.

Hopper Cars

Hopper car
A type of railroad freight car used to transport loose bulk commodities.

Hopper cars are generally constructed using metal with sides and ends that are fixed. These cars are used to transport dry bulk materials such as fertilizers, chemicals, salt, flours, and grains. The main hazard of a hopper car is the cargo itself.

Passenger Railcars

Passenger cars are railcars designed to carry people or provide specific services for passengers, such as riding, sleeping, dining, and luggage storage. Hazards associated with passenger cars begin with the number of persons onboard. Unlike freight trains, which have a small number of crewpersons, passenger trains may carry several hundred people all participating in a variety of activities. Older passenger cars have combustible interiors and are subject to fires from smoking materials and other ordinary combustibles found in the passenger riding areas.

Passenger cars also are equipped with cooling systems, air-conditioning systems, and electrical systems. All may contribute to or cause a fire problem and are of tactical concern for the firefighter.

Tank Railcars

Tank railcar
A tank mounted on a railroad frame with wheels designed to transport a variety of liquid products, having tank capacities ranging from a few hundred gallons to as much as 45,000 gallons (over 170 kiloliters).

Tank railcars vary widely in size and capacity. Even though tank cars can carry products that are as safe as purified water, a burning tank car fueled by a flammable liquid is a common image. The capacity of tank cars can range from a few hundred gallons to as much as 45,000 gallons (170.325 kiloliters) of product. Tank cars may have one or more compartments. They may be nonpressurized and carry a commodity such as gasoline, or pressurized and carry LPG. Tank cars may also be interconnected by piping, thus forming a tank train. Further, several 1-ton containers of chorine are commonly transported on a flat car, called a multiunit tank car.

The product and the type of tank car play an important role in determining the basic fire suppression strategy and tactics. For example, if flame is impinging on a pressurized railcar containing LPG and sufficient water supply is not available or cannot be applied to the point of impingement in a sufficient quantity to cool and reduce the internal tank pressure, then evacuation and a defensive or no attack mode should be undertaken. Figure 9-12 is an example of a defensive approach to a tank car fire.

Electric Locomotives

Third rail (wire)
The rail used in subways and the wire for overhead electric train power systems to provide electrical power to the train.

Electric locomotives are powered with electricity from a **third rail** or overhead wire carrying between 25,000 and 50,000 volts. Firefighters need to make sure no one touches or crosses over the powered third rail or wire as electrocution can occur.

Subway Rail Vehicles

Problems from a fire in a subway can be numerous. In a highly populated urban area there is always the possibility of a serious life loss as hundreds of persons could find themselves trapped on an underground train. The combination of poor lighting, intense heat, and dense smoke along with the feeling of confinement in an enclosed space combine to present a serious life threat to passengers,

Figure 9-12 *A defensive approach using unstaffed hose streams on a tank car.*

trainmen, and firefighters. Tunnels are poorly lit with the small emergency lights and if the third rail is still energized there is a serious life threat to all those working within the darkness of the tunnel.

Reaching the heart of these fires is challenging, and in some cases they are extremely difficult to extinguish. Hose lines must be stretched by hand unless the subway is equipped with a built-in standpipe system. A breathing apparatus is always needed, and supplying fresh breathing bottles complicates an already serious fire fight.

Fire Tactics and Strategy in Railed Transportation Equipment

Fires and other emergencies in railed transportation vehicles offer unique fire behavioral and other safety related problems. For example, they combine high-voltage electrical systems with diesel-powered/electric engines to drive the train. Large amounts of fuel are needed for the power system, as many are cross country transporters.

Rail systems are regulated by fire and safety requirements, which are not well understood by many fire departments, and because they haul a wide variety of cargo, they have many different configurations of vehicles and cargo-carrying units. Many of the cargos are large quantities of hazardous materials which, if mixed, can react explosively. Firefighters should first identify the hazard of the cargo or the mix of hazards and review the recommendations for evacuation distances found in the emergency guidebook.

If the fire can be fought safely, it should be approached downwind and with sufficient water supplies to protect the firefighters making the initial attack.

A review of the attack recommendations for tank cars carrying compressed flammable gases is always needed as they are subject to a BLEVE and special precautions need to be taken. Firefighters need to wear full protective equipment including SCBA as some of the materials carried can be extremely hazardous.

AIRCRAFT

The characteristics of aircraft fires and safety-related issues differ from other transportation equipment and structure fires because of the speed at which these fires develop and the intensity of heat generated from burning fuel. Speed is of concern in both the takeoff and landing phases as research indicates that about 80% of aviation accidents occur during these two phases. To further complicate the safety problem, when the aircraft is in the air there is no safe escape route; and passengers, once on the ground, find themselves in a long, narrow tube with very limited egress. Figure 9-13 shows a coordinated attack with aircraft fire fighting trucks.

The **Federal Aviation Administration (FAA)** regulates the amount and type of fire fighting systems located at airports using an indexing system. Under federal regulations, aircraft are sorted by their length into categories called indexes. They range from Index A, where the aircraft is less than 90 ft (27.43 m) in length, to Index E containing aircraft that are longer than 200 ft (60.96 m) in length. The FAA requires fire protection equipment based on the frequency of aircraft landings,

Federal Aviation Administration (FAA)
An agency of the DOT with authority to regulate and oversee all aspects of civil aviation in the United States.

Figure 9-13 *Aircraft fire fighting trucks make a coordinated attack.* Photo Courtesy of Lieutenant Dennis Leon, Dallas/Ft. Worth International Airport, Department of Public Safety.

Air bill

A document used to identify the hazardous materials being shipped by air. The captain is responsible for maintaining control over these papers.

type of aircraft using the runways, and the airport index. The regulations can be found in the *Code of Federal Regulations*, title 14, part 139 and a recent update by the FAA in advisory circular 150/5210-6B. Also, it is important for responders to know that each plane is required to have an **air bill** for each flight. The air bill identifies hazardous materials being shipped by air. The captain is responsible for maintaining control over these papers, located in the cockpit.

Aircraft Fuel

Aircraft fuels are identified by their ease of ignition. They have a wide range of flammability and are of special design as they are subject to changing altitudes, temperatures, outgassing of dissolved oxygen, and movement of the fuel from air turbulence.

There are a variety of jet fuels; however, the primary fuel used is identified as Jet A fuel. It is a kerosene grade fuel, which, because it has a higher ignition temperature than gasoline, is considered relatively safe during fueling operations. While the ignition temperature is higher than gasoline, this kerosene fuel once ignited is just as dangerous as gasoline fuel.

Aqueous film-forming foam (AFFF) is particularly suited for application to fires in aircraft fuel spills. It readily supplies a surface film over kerosene because of its fluidity. AFFF flows smoothly, helping to cover the fuel spill. This type of foam is compatible with most dry chemical agents, which also may be needed on aircraft fires. Firefighters must be aware, however, that AFFFs drain rapidly and provide less burn-back resistance compared to protein-based foams.

Aircraft fuels may also be released as a mist after the impact of an aircraft accident. The fuel mist is subjected to hot metal, disrupted electrical circuits, and other ignition sources. If the fuel is in a mist form, firefighters must be prepared for the possibility of an explosion or rapidly spreading fire. In some cases, foam can be used to hold vapors to a minimum and provide an escape avenue.

Research for developing antimisting additive is currently underway with a few promising developments. More testing and research is needed to fully develop an effective antimisting additive.

Another concern with aviation fuels is the need to make sure the aircraft is electrically grounded so as not to create the opportunity for a static spark that could ignite available vapors. Most airport runways are equipped with electrical grounding posts to prevent explosions from sparks that may occur.

Hydraulic Systems and Fluids

Aircraft like other transportation vehicles are equipped with hydraulic systems and hydraulic fluids that are operated under pressure. Always approach heated hydraulic cylinders with extreme caution. A fluid under pressure, when released, will spray as a mist. An ignition source can ignite the mist (vapor cloud), immediately resulting in fire or an explosion.

Oxygen Systems

Many jets have their own oxygen/air systems which are automatically put into action if cabin pressure drops. These systems can increase the flammability of materials in the passenger compartment including passengers' clothing. Firefighters need to understand the operation of these systems and know how to shut them down.

Electrical Systems

Most jet aircraft have special electrical generating units working at 24 volts. These systems provide energy for a number of the electrical power–driven motors that drive the wings, flaps, rudder, and other electrical systems within the aircraft. Some jet aircraft have hydraulically powered systems to provide a backup for the aircraft control systems. Aircraft fire and safety familiarization drills at the local airport can assist firefighters in understanding the various design features of the aircraft using their jurisdictional airport.

Anti-icing Fluids

Anti-icing fluids are used to keep ice off the wings and the moving parts of the wings and tail portion of the airship. These fluids are 85% alcohol and 15% glycerin. Some aircraft engines may also have alcohol/water injection systems. While not as great a hazard as other aircraft fuels, alcohol will burn with an almost invisible blue flame and may require greater amounts of water to dilute the fuel. In most aviation settings, the amounts of alcohol are limited.

Pressurized Cylinders

An aircraft has a number of different pressurized cylinders. For example, oxygen cylinders have pressure relief valves, while some pressurized cylinders may be used for hydraulic fluids, fire extinguishing systems, rain repellant systems, pneumatic systems, and other compressed gases. All of these cylinders have been known to explosively disintegrate during aircraft fire fighting operations.

Tire, Rim, and Wheel Assemblies

Large aircraft tires may have pressures up to and in excess of 200 psi. They are usually filled with nitrogen, an inert gas, to protect the tire from the tremendous amounts of heat generated during takeoffs and landings.

Because of the high pressures, these tires can explode with the force of a bomb when overpressurized, overheated, or damaged during a crash impact. Firefighters must take special caution when in close vicinity of these tires during fire fighting operations.

Escape Slides

The larger jet aircrafts use escape slides, which are automatically deployed and inflated within a number of seconds from the time the exit opens in the emergency mode. Firefighters arriving at a crash scene must prepare to assist passengers as they exit using these chutes, moving them quickly away from the aircraft body and wing areas.

Military Aircraft

Firefighters responding to military aircraft must be aware that explosives are used to eject the pilot seat and canopy in certain military aircraft. These devices eject the canopy and seat with great force. A preplanning tour of the nearest military base is recommended to familiarize personnel with the unique systems found on some military aircraft.

One of the greatest dangers of military aircraft is the wide variety of armaments located in and on the aircraft. Military aircraft may also transfer hazardous materials. Special training and experience are important in dealing with military aircraft fires and incidents.

Crash Scene Security

Aircraft emergencies attract the public and media attention almost immediately. It is important to establish an area surrounding the crash to allow the performance of emergency operations as well as to protect the scene for evidence. In order to do so, there are generally two perimeter zones established around an aircraft crash. The inner security perimeter is a minimum of 300 ft (98.43 km) from the outer limits of the wreckage and debris path. This provides room for the first responders to perform fire fighting, rescue, and medical functions as needed and any crime scene or crash scene investigation work that needs to be done. This inner security zone is surrounded by an outer security zone designed to contain spectators and members of the press. The second zone should extend from the inner zone to a minimum distance of at least another 300 ft (98.43 km). When first established, both areas must be large enough to open the space for all activities and keep the public safe. They both can be reduced later as the incident becomes under control.

Regulations for Aircraft

The FAA provides one set of basic regulations that are adhered to worldwide. The FAA has maintained that the most critical issue associated with fire safety is the rate of heat release. As the rate increases, the time available for safe egress from the vehicle decreases rapidly. The FAA subdivides aircraft incidents into three types of fire scenarios. The first fire scenario includes fires that occur after an air crash, where the aircraft fuel ignites, resulting in fire penetration into the fuselage or passenger area. The second and third fire situations involve fires occurring while an aircraft is in flight. These fires may occur in the passenger cabin or in the cockpit and spread rapidly in spite of being quickly detected. The third fire situation involves fire that remains undetected for a time, such as when a fire occurs in a hidden area (e.g., a restroom or cargo space).

The FAA conducted full-scale fire tests using actual aircraft for a variety of fast-progressing, inflight and postcrash fire scenarios. In each test, scientists assessed the release of heat, smoke, and combustion products. The test results indicate that new and improved fire performance of materials is needed. The test criterion was that the material in the passenger compartment and cockpit would allow passengers at least five minutes for escape from the aircraft. It was also understood that passenger egress could only start after the aircraft landed either by crashing or by pilot action. Some aircraft now have heat sensors with extinguishing systems in the cargo areas.

Aircraft Engines (On the Ground)

Fires in aircraft engines on the ground are generally not serious, because the fire attack can be made directly by ground units. There is a difference, however, depending on whether it is in a piston-driven engine or a turbine engine. If the fire is contained within the engine **nacelle,** the first effort should try to extinguish the fire by using the onboard extinguishing system. If this fails, then the fire will need to be extinguished using hose lines with fog nozzles to provide enough cooling to control the fire.

Jet (Turbine) Powered Aircraft Engine Fires

Fires in the combustion chamber of jet engines are best controlled if the engine can be kept turning over. Extreme caution should be observed, however, when working around a running jet engine. Personnel should never stand within 25 ft (7.6 m) of the front or the side, or directly to the rear of the engine outlets. The suction created by some engines is strong enough to draw in a 200-lb (90.8-kg) person.

Personnel should also stand clear of the turbine or rotation area. In the event of engine disintegration, this area will be the path of flying metal parts. The area to the rear of the exhaust outlet should also be avoided for a distance of at least 150 ft (45.6 m). Exhaust temperatures reach approximately 3,000°F (1,648.9°C) at the outlet.

Extinguishment of fires that are outside the combustion chamber but within the engine nacelle would first be attempted by using the aircraft's built-in extinguishing system by the pilot or crew. If this is unsuccessful then carbon dioxide or dry chemicals should be tried; however, these extinguishers should not be used if magnesium or titanium metals are involved. Under these circumstances,

Nacelle

The covering or cowling that protects the engine of the aircraft.

it is best to allow the fire to burn itself out. Foam or water spray should be used to keep the nacelle and surrounding exposed parts of the aircraft cool.

Wheel Fires

Potential wheel fires should be approached with caution. The responding fire apparatus should be parked within effective fire fighting distance from the aircraft but never to the side of the aircraft or in line with the wheel's axle. If the tire should explode, tire debris is generally thrown to the sides but not to the front or rear. Consequently, if it is necessary for firefighters to approach the aircraft, it should be done from the front or rear.

Smoke around the brake drums and a tire does not mean that the wheel is on fire as the overheating of brakes occurs often. If the brakes are overheated, they should be allowed to cool by air only using a smoke ejector to help reduce the cooling time. Do not use water for cooling, as the rapid reduction of heat on the wheels may result in an explosion. Carbon dioxide extinguishers are frequently used for cooling wheels and can be used if there are actual flames in the wheel. However, a dry chemical extinguisher is preferred and recommended because it is less likely to chill the metal in the wheel parts.

Water can be used only when a dry chemical extinguisher is not available; however, it should be used with caution. Firefighters should protect themselves from a possible explosion by using the fire apparatus as a shield. The water should be applied in a fine spray and in short bursts of five to ten seconds. At least thirty seconds should elapse between bursts. Water should be used only as long as flames are visible.

Strategy and Tactics in Aircraft

Aircraft fires and accidents present unique problems considering that passengers are essentially trapped in an aluminum tube surrounded by flammable fuel. The fuel, once released, can quickly invade the passenger compartment and engulf the entire aircraft in flames. Quick departure from the aircraft or evacuation by the passengers and rapid extinguishment actions by firefighters are absolutely necessary to save lives.

Rapid intervention vehicle (RIV)
The aircraft RIV is one example designed to rapidly deploy firefighters.

The size-up of these incidents must consider the escape and extinguishment of the fire. If the fire cannot be extinguished, then a relatively safe escape route must be provided by confinement of the fire using a (aircraft) **rapid intervention vehicle (RIV).** An aircraft RIV is a large pumper specifically designed for aircraft fire fighting. It carries large amounts of water and foam. Some of these vehicles have two engines. One engine propels the vehicle quickly, while the second engine powers the pump allowing a fast-moving attack while pumping foam or water.

These RIVs have a roof-mounted monitor (foam application device), which can be operated from the cab, thus discharging foam while advancing on the fire

Figure 9-14 *Aircraft rescue fire fighting vehicle with a firefighter in specialized protective clothing.*

(see Figure 9-14). The foam provides a cooling and smothering action and blankets the vapors from the leaking fuel to prevent ignition. This quick application of foam can provide the necessary confinement of the fire to allow passengers to escape. Some units carry large, dry powder extinguishers for use on magnesium and other combustible metal fires which may occur.

Firefighters have special temperature-resistant protective clothing designed to withstand the high temperatures generated by aviation fuels. Special tools are used to penetrate the aircraft skin to provide application of foam inside the fuselage when needed. Specialized training and equipment are needed for aircraft fire fighting.

BOATS

The possibility of a small boat fire exists in any fire department's jurisdiction that has a body of water of any size. In most cases, small boats are brought to the water on a trailer, but some boats may be docked in a boat marina. Boat marinas will vary in size from those docking only a few boats to those storing a large number. The method of storage usually consists of one or more floating docks extending outward into the body of water.

Many of the small boats in the marina are protected from the weather during nonuse by canvas covers that extend from one end of the boat to the other. This large canvas cover presents a potential fire problem as fire can rapidly spread in the event that one of the boats becomes involved in a fire. Firefighters must preplan for this type of runaway fire and develop plans to reduce the risk of destruction of all boats stored in the marina.

Getting water to a boat fire may be a problem if the boat is tied up to a dock far from the shoreline and water is not available. Again, during the fire preplanning, firefighters should develop alternative methods to obtain water. Some of the larger boat marinas provide fire hydrants adjacent to the walkways. They should be tested periodically. At other marinas it will be necessary to **draft** the supply of hose lines. If engines cannot be positioned close enough to draft, a siphon ejector may be used.

Fires in the cabin or superstructure area should be attacked using foam or water spray streams. A single 1.5-inch (37-mm) or 1.75-inch (45-mm) line will normally be sufficient for a fire in a small boat. Two lines will normally be required if the fire is sizable or in a small yacht. A fire in a small yacht should be attacked from both sides so as not to push the fire onto an adjacent docked boat.

Most fires on small boats start near the engine or in the bottom area (**bilge**) of the boat. This area in some cases traps flammable vapors, which may be ignited when the engine is started and the fire usually is initiated with an explosion. A carbon dioxide or dry chemical extinguisher is very effective if the fire is fairly well contained under the floor decking and has not advanced too far.

Water in the form of fog or spray streams is also effective; however, it presents the possibility of sinking the boat and spreading burning fuel onto the adjacent waterway. Thought should always be given to extending protection to boats docked nearby.

Regardless of the methods used to extinguish the initial fire, there will probably be glowing materials in the combustible furnishings inside the boat. These items need to be thoroughly overhauled. There is also the possibility of considerable amounts of unburned gasoline in the bilge which will have to be siphoned into an approved container or it may be necessary to request a flammable liquid vacuum truck if the spill is extensive. Caution should be taken to ensure that the wiring is disconnected from the batteries and shorelines prior to commencing these operations. It may be necessary to pump out (or dewater) the boat to keep it from sinking if too much water is used.

Occasionally it will be found that a small boat on fire has been set adrift prior to the arrival of the fire department to keep it from damaging adjacent boats. In such cases the boat should be secured prior to attacking the fire with hose lines. If this is not done, hose stream application may drive the boat into adjacent exposures.

Ship Fires

In the United States, regulations are established by the federal government and enforced by the U.S. Coast Guard (USCG). There may also be local and state laws, but in most cases these are usually in accordance with federal regulations.

Draft

The process of moving or drawing water away from a static source of water by a pump.

Bilge

The bottom area of a boat or ship. This area of the ship or boat can accumulate flammable vapor and flammable liquids.

Safety of Life at Sea (SOLAS)

An international treaty containing minimum standards of fire and related safety issues for ships on international voyages.

International Maritime Organization (IMO)

An agency of the United Nations dealing with maritime issues. It is responsible for maintaining the SOLAS treaty.

Bulkhead

A main wall or supporting structure generally made of steel.

Dangerous cargo manifest

Documents used to identify the hazardous cargo of a marine vessel.

For international waters, a treaty was approved by a convention of concerned nations, titled the **Safety of Life at Sea (SOLAS).** This international treaty contains minimum fire standards and related safety issues for ships on international voyages. Because the United States is a signatory to the SOLAS treaty, U.S. registered (flag) ships using international waters must comply with its provisions.

Another agency concerned with fires and safety regulations at sea is the **International Maritime Organization (IMO).** It is an agency of the United Nations specializing in maritime affairs. IMO consists of national governments with maritime activities. It has the responsibility for maintaining the SOLAS treaty and has issued a number of fire test methods (IMO resolutions), which are referenced in the SOLAS treaty.

Fires in the Hold of a Ship

Fires aboard ships are greatly influenced by the conductivity of the steel construction. The conductivity and transmission of heat by the steel can spread fire among compartments, making fires more difficult to control than most structure fires. In addition to the conductivity of the steel hull, other problems include the generation of smoke and heat. Ship cargo holds are confined spaces where smoke and heat can build, and without proper ventilation this combination of intense heat and thick, toxic smoke makes fire extinguishment dangerous and difficult for firefighters. Ships have **bulkheads** which are the main wall or supporting structure generally made of steel. Some bulkheads are watertight and fire resistive. They can be designed and constructed to prevent the spread of water and fire if the ship has been damaged.

As with all fires, size-up requires the first arriving firefighter to determine the location of the fire, find out what is burning, and determine the extent of the fire. Cargo ships are required to have a **dangerous cargo manifest,** or a listing of what is being carried on the ship. It is important to find out if any hazardous materials are being carried and, if so, in what cargo holds they are being stored and the hazards they present.

The firefighter needs to determine if the ship is equipped with a CO_2 system or steam fire extinguishing system and if the system is operable. If the system is available, it can be used after a review of the cargo manifest, the extent of the fire, and the ability of the system to hold or contain the fire. Always check the compatibility of the extinguishing system with the materials being impacted by the fire.

If available and compatible with the cargo, foam of medium or high expansion (20 to 1,000 times) may be used to fill enclosures such as holds of ships where fire is difficult or impossible to reach. Here, foam acts to halt convection currents and reduce access to air for combustion. The water content of the foam also cools and diminishes the available oxygen by steam displacement.

It is important to remember that oxidizers provide their own oxygen when heated. Therefore, a fire involving an oxidizer should not be extinguished by filling the hold with CO_2 or trying to exclude the oxygen by the introduction of steam. It is recommended that students review the report of the fire and explosion of the *S.S. Grand Camp* in Texas City, Texas, in 1947. The review is available at https://www.usfa.dhs.gov/applications/publications/.

One problem using water to extinguish fires on a ship is the possibility of capsizing the ship as the fire fighting water may be absorbed by the cargo or the added water can upset the ship flotation balance and impact the stability of the ship. The responsibility for the stability of the ship is not with the fire department but the ship's officer. So it is important to keep in close communication with the ship's officers if the stability of the ship is being threatened due to the amount of water being pumped into the hold.

Tanker Ships

Tanker ships present a fire problem that differs from the cargo ship as the tanks or cargo area contains flammable liquids, which may be extinguished by suffocating the fire. Generally one of two types of fire extinguishing systems is installed. The two systems are either a CO_2 or a steam system. Both will work on flammable liquids if all cargo tank openings can be closed to reduce the inflow of additional air (oxygen). Once the ship's fire extinguishing system is in service, firefighters need to verify that the system is operating properly.

In some cases, it may not be possible to close all openings and foam will be the only effective agent available. Water can be used sparingly, spraying to precool the liquid prior to the application of the foam. Foam should not be applied if materials such as asphalt or tar are involved as this could result in a steam explosion.

Fire Tactics and Strategy on Boats and Ships

Marine fires have many of the same hazards that complicate fire fighting activities as other classifications of transportation vehicles. Like aircraft, when on the water and carrying passengers, safe evacuation is important, as boat and ship fires can move quickly and produce a great deal of deadly smoke. Ships carry a wide variety of cargo, some of which may be hazardous. The electrical, fuel, and ventilation systems can malfunction, creating either a fire or other safety problem.

Small boats are constructed of wood, fiberglass, or aluminum, all of which can burn or create safety problems for those onboard. Larger ships are mostly constructed of steel with large cargo spaces, which are poorly ventilated. The steel transmits heat, which can spread the fire to other cargo compartments.

In sizing up these fires, firefighters need to obtain the cargo manifest to determine the nature and problems presented by the cargo, and the type and condition of the onboard fire fighting system, if one is available. Full protective clothing with SCBA and an extra supply of air bottles will be needed if a large fire occurs. Fire preplanning and close coordination with the harbormaster is highly recommended to prepare firefighters for marine fires. In extreme situations, it may be preferable to sink the boat or ship in shallow water and then refloat it when the fire is extinguished. Figure 9-15 shows a cargo ship in port.

Figure 9-15 *A cargo ship in port.*

SUMMARY

A review of the average annual transportation fires shows that fires and the related safety issues in transportation vehicles represent a big problem for firefighters in the United States. Each year approximately 400,000 fires occur in vehicles, almost two-thirds of the average number of structure fires.

Each category of transportation fires presents specific fire and safety-related problems.

These problems are found in the propulsion or power system, the fuel, electrical and hydraulic systems, and cargo areas. These locations present situations where uncontrolled fires and hazardous materials can create a safety hazard.

For the sake of safety, it is important to preplan and review the design and specifications of transportation vehicles that firefighters may encounter during a fire.

KEY TERMS

Air bill A document used to identify the hazardous materials being shipped by air. The captain is responsible for maintaining control over these papers.

Bilge The bottom area of a boat or ship. This area of the ship or boat can accumulate flammable vapor and flammable liquids.

Bill of lading Documents indicating the amount and type of cargo or freight being transported.

Boxcar A railroad car that is enclosed and used to carry general freight.

Bulkhead A main wall or supporting structure generally made of steel.

Compressed natural gas (CNG) Natural gas that is compressed and contained within a pressurized container. It provides the fuel to power some of the new hybrid vehicles and many public buses.

Consist (way) bill Documents used to identify the hazardous cargo of a rail transportation vehicle. The driver of the vehicle has the responsibility to maintain these documents.

Cribbing   Blocks of wood or prefabricated plastic blocks used to lift a vehicle's frame off the wheels and stabilize the vehicle.

Dangerous cargo manifest   Documents used to identify the hazardous cargo of a marine vessel. The captain or master of the ship has the responsibility to maintain these papers.

Draft   The process of moving or drawing water away from a static source of water by a pump.

Federal Aviation Administration (FAA)   An agency of the DOT with authority to regulate and oversee all aspects of civil aviation in the United States.

Flatcar   A railcar that is open to the elements.

Gondola car   Railcars designed with flat bottoms and four walls that can carry a variety of goods. They may have a cover for the cargo being carried. A few are constructed of wood, but most often they are constructed of steel.

Hopper car   A type of railroad freight car used to transport loose bulk commodities.

Hybrid passenger vehicles   A term used to describe vehicles powered by a variety of fuels and powering systems. Hydrogen, compressed natural and liquefied petroleum gases, as well as preheated vegetable oil can be used. The powering systems in some vehicles use higher voltage systems, up to 600 volts. Some vehicles combine gas turbines and a battery system. Most are experimental and a number have resulted in fires. New hydrogen fueling stations are being designed and some are already built. Firefighters need to stay on top of the research and development of products in this area.

Inflator   A pressurized container that holds gas that is released upon impact, filling the curtain or bag which restrains passengers within the seating area. The container in the newer hybrid cars can contain 3,000 to 4,000 psi.

Intermodal railcar   Transportation of freight in a container or vehicle, using multiple modes of transportation (rail, ship, and truck), without any handling of the freight itself when changing modes.

International Maritime Organization (IMO)   An agency of the United Nations dealing with maritime issues. It is responsible for maintaining the SOLAS treaty.

Nacelle   The covering or cowling that protects the engine of the aircraft.

National Highway Traffic Safety Administration (NHTSA)   A group that developed a small-scale regulatory fire test for vehicle interior materials in the late 1960s and has continued applying the same test today.

Rapid intervention vehicle (RIV)   The aircraft RIV is one example designed to rapidly deploy firefighters. For example, Los Angeles County has smaller and faster pumpers that carry limited water. The unit is able to respond rapidly in an effort to extinguish fires before they grow in size.

Safety of Life at Sea (SOLAS)   An international treaty containing minimum standards of fire and related safety issues for ships on international voyages. The United States is a signatory to the SOLAS treaty; therefore, U.S. registered (flag) ships using international waters must comply with its provisions.

Side impact curtain (SIC)   A new vehicle passenger restraint system that deploys more quickly and forcefully from the side because of the shorter distance to the passenger.

Supplemental restraint system (SRS)   An additional passenger restraint system, some of which are placed in the vehicle roof, dash, or in a roof pillar.

Tank railcar   A tank mounted on a railroad frame with wheels designed to transport a variety of liquid products, having tank capacities ranging from a few hundred gallons to as much as 45,000 gallons (over 170 kiloliters).

Third rail (wire)   The rail used in subways and the wire for overhead electric train power systems to provide electrical power to the train.

REVIEW QUESTIONS

1. Most firefighters as well as U.S. citizens do not realize the high number of transportation fires that occur or the losses that result from them. Review Table 9-1 and provide explanations for these facts.

2. Describe the reasons why today's passenger car fires are more difficult to extinguish.

3. Describe the types of fuels that firefighters may encounter in the new hybrid passenger vehicles.

4. Describe some of the fire problems that may be encountered on fires involving motor homes, buses, and recreational vehicles.

5. Why are ship fires difficult to fight? Include some of the fire behavior problems that may be encountered on a ship.

REFERENCES

Association of American Railroads. "Hazardous Materials Regulations of the Department of Transportation," Bureau of Explosive Tariff No. BOE-60000 (Washington, DC: Bureau of Explosives, Association of American Railroads).

Burnett, John. (1998). "Fire Safety Concerns for Rail Rapid Transit Systems," *Fire Safety Journal* 8(1), 103–106.

Federal Emergency Management Agency and Charles Jennings. (1990). "Gasoline Tanker Incidents in Chicago, Illinois and Fairfax County, Virginia: Case Studies in Hazardous Materials Planning," Technical Report 032.

Gesell, Laurence E. (1999). *The Administration of Public Airports* (Chandler, AZ: Coast Aire Publication).

Gustin, Bill. (1996, March). "New Fire Tactics for New Car Fires," *Fire Engineering*, 62–67.

———. (1997, August). "Expect the Unexpected in Vehicle Fires," *Fire Engineering*, 87–94.

National Fire Protection Association. (2004). *The Research Advisory Council on Fire and Transportation Vehicles* (Quincy, MA: National Fire Protection Research Foundation).

———. (2008). *Fire Protection Handbook* (20th ed). National Fire Protection Association, 6–270.

Shaw, Ron. (2005, April). "New Auto Safety Technology," *Fire Engineering*, 78–82.

Smith, D. A. (1984). "Some Aspects of Fire Safety Design on Railways," *Fire and Materials* 8(1), 6–9.

St. Louis, Ed and Steve Wilder. (1999, June). "Train Disasters Test the Fire Service: Tragedy on the City of New Orleans," *Fire Engineering*, 61–71.

U.S. Coast Guard. (2004). *Amendments to the 1994 Safety at Sea (SOLAS) Treaty,* Navigation and Vessel Inspection Circular 4-04.

U.S. Department of Transportation. (2008). *Emergency Response Guidebook* (Washington, DC: Government Printing Office). Copies can be downloaded at the Environmental Protection Agency website, http://www.epa.gov/epahome/abouepa.htm.

U.S. Fire Administration. (2004). *Transportation Fires, The Research Advisory Council on Fire and Transportation Vehicles* (Quincy, MA: National Fire Protection Research Foundation).

Chapter 10

HAZARDOUS MATERIALS AND WARNING SYSTEMS

Learning Objectives

Upon completion of this chapter, you should be able to:

■ Describe the U.S. Department of Transportation hazardous materials warning system, including its advantages and disadvantages.

■ Describe the National Fire Protection Association standard 704 warning system, including its advantages and disadvantages.

■ Explain the requirements, purposes, and value of Materials Safety Data Sheets to first responders.

■ Explain the types of information available for first responders contained in the 2008 *Emergency Response Guidebook.*

■ Describe the responsibilities and duties of state and local emergency planning committees and how their plans and documents assist first responders with information.

■ Explain the issues that make weapons of mass destruction incidents complex and the reasons for the development of the National Incident Management System.

INTRODUCTION

This chapter provides an overview of the systems designed to warn first responders of the dangers of hazardous materials, reviews legislative actions designed to control hazardous materials, and provides first responders with an overview of regulations enacted for those who are likely to encounter accidentally or intentionally released hazardous materials.

Both private and governmental agencies have developed a number of warning and information systems. This text will look at the three major systems used by first responders. Stressing how vital these warning systems are to first responders and the need for their continuation and improvement is the goal of the International Association of Fire Chiefs (IAFC). A 2005 survey of fire chiefs found "98% of those chiefs responding indicated that the warning systems were critical."

A final topic in the chapter is a relatively new threat to first responders, weapons of mass destruction (WMD), and their impact on how emergency management, planning, and preparedness is managed and organized.

HAZARDOUS MATERIALS AND FIRST RESPONDERS

In the 1940s, new types of emergency incidents began to emerge in the United States. These incidents were linked with the behavior of certain chemicals, petroleum, and nuclear products that began to appear in the marketplace as products of convenience. The basic building blocks for these products were called **hazardous materials.** They came to the attention of first responders when they were misused or involved in unintended fires or mishaps. Today many of these hazardous materials have become a vital part of culture as they have made it possible to enjoy many conveniences of modern living. Eliminating these hazardous materials from our society is not possible; therefore, it is vital that their usage is as safe as possible by effectively controlling their handling, transportation, and storage.

In the 1940s, only a few hazardous products were available to the public. By the early 1970s, demand for them grew significantly and, with the increase in demand, more problems began to surface. It became apparent then that something needed to be done to control the handling of these substances.

At the national level, the federal government began to address the problem of abandoned hazardous wastes. The concern resulted in Congress enacting the Resource Conservation and Recovery Act of 1976, which was followed by the Comprehensive Environmental Response Compensation and Liability Act (CERCLA) in 1980 and the Superfund Amendment and Reauthorization Act (SARA) in 1986. These three important legislative documents were signed by

Hazardous materials
Any material in any form or quantity that poses an unreasonable risk to safety and health and property when transported in commerce.

Hazardous substance

Any substance posing a threat to waterways and the environment when released.

National Response Team (NRT)

A team of fourteen federal agencies assembled to assist state and local agencies with hazardous materials incidents and the pre-incident planning for these incidents.

Presidents Carter and Reagan and they combine to form the basis of the superfund program that provides funding to mitigate the release of **hazardous substances.**

Other legislation and federal actions have resulted in the National Oil and Hazardous Pollution Contingency Plan (NCP). This plan is designed to assist local communities in preventing and mitigating harm from hazardous materials releases. Local first responders have been assigned duties and responsibilities under this plan. A review of the plan and its requirements should be conducted by all local agencies. In an effort to provide better federal agency coordination, the **National Response Team (NRT)** was formed, bringing together fourteen federal agencies to provide assistance to states with their emergency response planning as well as assistance during major hazardous materials emergencies.

As co-chairs of the NRT, the Environmental Protection Agency (EPA) and the U.S. Coast Guard play major roles in environmental protection. These two agencies provide waterway protection. The EPA has inland water responsibility, while the Coast Guard handles coastal waters and specifically designated federal navigable waterways such as Lake Michigan. Together these organizations and related laws are responsible for implementation of a federal program for preventing, mitigating, and responding to releases of hazardous substances that might threaten human health and the environment.

In 1998, in response to growing concerns over acts and threats of terrorism, President Clinton issued presidential decision directive 62 (PDD 62). The purpose of the directive was to "create a new and more systematic approach . . . to achieve meeting the threats of terrorism in the 21st century" (p. 12).

The directive established a national coordination for security, infrastructure, protection, and counterterrorism. As a result, a nationwide effort is being made to identify potential targets, develop response and consequence plans, review the current state of personal protective equipment, and train first responders of the 120 largest U.S. cities regarding the hazards, response procedures, and equipment required to respond to an incident involving a chemical or biological agent. When the personnel in these cities are trained, others in remaining U.S. cities will require the same training to ensure nationwide standardization.

Materials Safety Data Sheets (MSDS)

A required worksheet prepared by manufacturers and marketers to provide workers with detailed information about the properties of a commercial product that contains hazardous substances.

These actions and related legislation are important to the fire service as first responders, because the legislation is fairly comprehensive and provides some funding provisions. There are provisions for funding of training, mitigation, and enforcement of regulations specific to hazardous materials incidents. In addition, there is a requirement to notify first responders of an existing hazard being stored on a property or being transported. This legislation requires that **Materials Safety Data Sheets (MSDS)** be provided on-site where the materials are stored. MSDS contain vital information for first responders. They are explained in detail under the Emergency Planning and Community Right-to-Know Act.

TWO MAJOR WARNING SYSTEMS

Two warning systems have the specific purpose or function to provide a heads-up warning for first responders. They are the hazardous material placard utilized in transportation and the color-coded hazardous warning system for fixed facilities.

The two organizations responsible for the warning systems are:

- U.S. Department of Transportation (DOT), and
- National Fire Protection Association (NFPA).

U.S. Department of Transportation System

Department of Transportation (DOT)
A federal department responsible for issuing standards and regulations relating to the transportation of hazardous materials nationwide.

The **Department of Transportation (DOT)** is the federal agency responsible for the regulations for the safe transportation of hazardous materials in various transportation modes. The DOT has established a required warning system for use on vehicles transporting hazardous materials on the highways. This system divides hazardous materials into nine major classifications of hazards, and then into further subdivisions to assist in specific identification of a substance.

The DOT's placard system is designed and used for vehicles that transport these materials over the highways. Because these vehicles transport across state lines, a nationwide standardized warning system is needed rather than each state having its own system. Recently, the DOT, in cooperation with the United Nations, developed an international classification system that identifies hazardous materials in transit with color-coded symbolic warning labels.

Emergency Response Guidebook (ERG)
A booklet that contains the chemical names and the commonly used names for most hazardous chemicals.

The DOT system uses identification numbers to categorize chemicals. This information can be found in the *Emergency Response Guidebook* **(ERG)**. This manual is provided by the DOT for every first response emergency apparatus nationwide, and is also distributed in Canada and to some South American countries. It contains chemical information sorted by a chemical identification number and a chemical name (both scientific and trade names). The ERG also includes information about placards and other critical reference information such as isolation distances and tank car identification methods. The ERG's cross-reference system is helpful, for example, if information from any one of the identification methods is known, then other methods are cross-referenced for verification and additional information.

The DOT warning label system combines four different ways to notify the first responder. The placard has a picture of the hazard at the top of the triangle. The picture depicts the hazard and the placard color identifies the hazard class. In the middle of the placard, the hazard is identified in plain English and at the bottom of the placard the hazard class and division number are displayed.

As previously mentioned, the DOT system divides the hazardous materials into nine major classes. Each of the nine classes is further subcategorized for specific material descriptions. For example, an explosive that is Class 1.1 has the most potential for damage or is more dangerous than a material classified as a Class 1.6 explosive, which is very insensitive to detonation. The nine classes identified by the DOT system are listed in Table 10-1.

Table 10-1 *DOT nine classes of hazards.*

Hazard Class	Product Example
Class 1—Explosives	
1.1 Explosives with a mass explosion hazard	Black powder
1.2 Explosives with a projection hazard	Detonating cord
1.3 Explosives with significant fire hazard	Propellant explosives
1.4 Explosives with no significant blast hazard	Practice ammunition
1.5 Very insensitive explosives; blasting agents	Prilled ammonium nitrate
1.6 Extremely insensitive detonating substances	Fertilizer–fuel oil mixtures
Class 2—Gases	
2.1 Flammable gas	Propane
2.2 Nonflammable gas	Anhydrous ammonia
2.3 Poisonous gas by inhalation	Phosgene
2.4 Corrosive gases (Canada)	Hydrogen bromide; anhydrous
Class 3—Flammable liquids, combustible liquids	Gasoline, kerosene
No divisions	
Class 4—Flammable solids, spontaneously combustible materials and dangerous when wet materials	
4.1 Flammable solids	Magnesium
4.2 Spontaneously combustible materials	Phosphorus
4.3 Dangerous when wet materials	Calcium carbide
Class 5—Oxidizers and organic peroxides	
5.1 Oxidizers	Ammonium nitrate
5.2 Organic peroxides	Ethyl ketone peroxide
Class 6—Poisonous materials	
6.1 Poisonous	Arsenic
6.2 Infectious substances (etiological agent)	Rabies, HIV, hepatitis B
Class 7—Radioactive materials	Radio active isotopes
No divisions	
Class 8—Corrosive	Caustic soda
No divisions	
Class 9—Miscellaneous hazardous materials	
9.1 Miscellaneous (Canada only)	PCBs, molten sulfur
9.2 Environmental hazardous substances (Canada only)	PCB, asbestos
9.3 Dangerous wastes (Canada only)	Fumaric acid

Source: *Emergency Response Guidebook* (U.S. Department of Transportation, 2008).

DOT System Advantages and Disadvantages

The ERG states that its purpose is to provide "a primary guide to aid first responders to quickly identify the specific or generic hazards of the material(s) involved in the incident . . . and to protect themselves and the public during the initial response phase."

Even though the ERG does not provide specific information of every hazardous material, this disadvantage may be overcome by checking the bill of lading or the MSDS, which should be attached to the transporting agency's documentation. Once the specific materials are identified, the ERG contains basic recommendations regarding handling, mitigating, and evacuation procedures.

A helpful feature of the ERG is the inclusion of recommended isolation and evacuation distances and response guidelines with special instructions for specific products. Phone numbers and other contract information are also included. First responders must also remember that exceptions to the transportation regulations may allow certain materials to be shipped without labels, placards, or markings. Such exceptions occur often in the transportation industry.

National Fire Protection Association System

The National Fire Protection Association (NFPA) is a nonprofit corporation with the mission of reducing the impact and burden of fire and other hazards in communities. One of the many functions of the NFPA is the development of consensus-based codes and standards. The NFPA has developed 300 codes and standards covering many buildings, processes, services, and equipment installations in the United States. The warning method used for materials stored in buildings is the **National Fire Protection Association (NFPA) 704 system.** The NFPA 704 standard for placards is vital in providing a warning to first responding units.

The NFPA 704 placard system is used as a method to warn emergency responders that hazardous materials are stored in storage tanks, small containers, and fixed facilities. A color and numbering system provides first responders with a warning that hazardous materials are stored or handled on the premises. Again, this requirement does *not* provide the first responder with specific information regarding the substance, but similar to the DOT requirement, the building owner is required to have MSDS containing specific material information on hand. Figure 10-1 is an example of a NFPA 704 symbol.

The placards are diamond-shaped figures divided into four quadrants, each of which is color coded corresponding to one of three hazards (blue for health hazards, red for flammability hazards, and yellow for chemical reactivity hazards) and numbered 0 to 4 (with 4 indicating the greatest degree of hazard) to designate the degree of the relevant hazard. The fourth (or bottom) quadrant is used to warn of special hazards such as water reactivity (W) or that the material is an oxidizer (O). Other hazard signs include the tri-foil, which is used to indicate a radiation hazard, ALK for alkalis, and CORR for corrosives.

Figure 10-1 *Example of a NFPA 704 hazard warning placard used for fixed facilities.*

Figure 10-2 *NFPA placard.*

Categories and definitions for the NFPA 704 placard system are provided in the textbox on page 246.

Figure 10-2 tells us that the material is a serious health hazard, but does not present a flammability hazard or a serious reactivity hazard. The material also is identified as an oxidizer.

NFPA 704 System Advantages and Disadvantages

Like the DOT warning system, the NFPA placard system also provides a quick, visual system by using a color coded placard that is displayed on the exterior of the building. It warns the first responder to take notice of hazardous materials stored in the building. However, a disadvantage is that many fixed plant facilities have more than one hazardous product on hand. While each chemical may have an NFPA placard on its packing carton, the placard located on the exterior of the business summarizes the highest hazard of each categorized chemical. For example, a building used to store chlorine, ammonia, methyl ethyl ketone, and acetylene has the following information.

- Chlorine is rated individually as 4 (Health), 0 (Flammability), and 0 (Reactivity) or 4-0-0.

- Ammonia (anhydrous) is rated individually as 3 (Health), 1 (Flammability), and 0 (Reactivity) or 3-1-0.

- Methyl ethyl ketone is rated individually as 1 (Health), 3 (Flammability), and 0 (Reactivity) or 1-3-0.

- Acetylene is rated individually as 0 (Health), 4 (Flammability), and 2 (Reactivity) or 0-4-2.

In this case, the NFPA placard on the front door would read 4 (Health), 4 (Flammability), and 2 (Reactivity). The numbers are not added up, but the most

NFPA Placard Identification System

Flammability

4 Flammable gases, volatile liquids, pyrophoric materials
3 Ignites at room temperatures
2 Ignites when slightly heated
1 Needs to be preheated to burn
0 Will not burn

Reactivity

4 Can detonate or explode at normal conditions
3 Can detonate or explode if strong initiating source is used
2 Violent chemical change if temperature and pressure are elevated
1 Unstable if heated
0 Normally stable

Health

4 Severe health hazard
3 Serious health hazard
2 Moderate hazard
1 Slight hazard
0 No hazard

dangerous one (the highest number) of each hazard is used. Although this is not a distinct disadvantage, it can lead to confusion. An actual chemical that is categorized as 4-4-2 is not even present although the warning placard would indicate such (Gantt, 2009).

Figure 10-3 shows a university warehouse. The hazardous materials stored inside may be surprising.

Another important point to remember is that the hazardous materials placard system is not chemical specific. No chemical identification system can accurately evaluate the effects of one chemical combining with another or the possible effects of a combination of unknown amounts of several chemicals.

Another weakness in this system is that it is not possible to determine the exact chemicals that are present, and it is not possible to know how many chemicals are in the building or where they are located. While this situation may not provide the emergency responders with all of the needed information, it is a good place to start. To find out specific preincident planning information, the

Figure 10-3 *NFPA 704 placard on a warehouse door.*

Emergency Planning and Community Right-to-Know requirements support the system of informing the first responders.

Emergency Planning and Community Right-to-Know Act (EPCRA)
An act that provides the public and local governments with information concerning potential chemical hazards present in their communities.

EMERGENCY PLANNING AND COMMUNITY RIGHT-TO-KNOW ACT

Under the 1986 **Emergency Planning and Community Right-to-Know Act (EPCRA),** each state governor designates a **state emergency response committee (SERC).** The SERCs in turn designate local emergency planning districts and then appoint **local emergency planning committees (LEPCs)** for each

State emergency response committee (SERC)

This committee is responsible for the coordination of emergency plans, which are developed by the LEPCs.

Local emergency planning committee (LEPC)

A committee that is responsible for developing and maintaining a local emergency response plan that ensures quick and effective response to chemical emergencies.

district. The SERC supervises and coordinates the activities of the LEPC, establishing procedures for receiving and processing public requests for information collected under the terms of the regulations, and reviews local emergency response plans.

The membership of the LEPC must include local officials such as police, fire, civil defense, public health, transportation, and environmental professionals as well as representatives of facilities subject to the emergency planning requirements, community groups, and the media. The LEPCs have the responsibility of developing an emergency response plan for their jurisdiction, ensuring an annual review, and providing information to the community about the chemicals stored within the community.

The four major area requirements to protect the community and first responders from hazardous materials releases are:

- Emergency planning;
- Emergency release notification;
- Hazardous chemical storage reporting; and
- Toxic chemical release inventory.

The emergency planning requirement calls for an emergency plan to be developed to handle a hazardous materials incident. Included in the plan are emergency response procedures, equipment and facilities, evacuation plans (if needed), and training for first responders.

The law requires that the facilities housing the hazardous materials provide emergency notification of a release by dialing 911. There are specific information requirements that must be provided to the 911 centers or telephone operator. The law also requires the facilities to maintain MSDS and a record of an estimated amount of hazardous materials released. The MSDS requirement is used to supplement both the DOT and NFPA warning systems and provides more specific information for product end users, handlers, and first responders. Table 10-2 details the information requirements for the MSDS.

MSDS required by the Occupational Health and Safety Administration are depicted in Figure 10-4.

The required MSDS provide the first responder with supplemental information that is more in depth than the initial information provided by either DOT or NFPA warning systems.

OTHER INFORMATION SOURCES

Chemical Transportation Emergency Center (CHEMTREC)

In some cases, even the MSDS will not provide the emergency responders with enough specific information on a chemical or compound and how to deal with it

Table 10-2   *Requirements for MSDS forms.*

Information	Note
Chemical product and company identification	The name the chemical company uses to identify the product is used near the top of the MSDS. Information related to the manufacturer, such as the address and other contact information, is provided near the top of the first sheet. The MSDS sheet is required to list a 24-hour emergency contact number, although in most cases this number is Chemtrec.
Chemical composition	The ingredients of the chemical are listed here. If the mixture is a trade secret, there is a provision to exclude this information. If the information is needed for medical reasons, the manufacturer is to provide this information to a physician. In some cases the true identity of the material may never be known, and only the hazards may be provided.
Hazards identification	In this section an emergency overview is provided, for both an employer and emergency responders. This section is lengthy and includes potential health effects, first aid, and firefighting measures.

The exposure levels are typically included with the health effects, but may be listed separately. New to MSDS will be a section on accidental releases, which will be beneficial to emergency responders. Handling and storage considerations and engineering controls, including proper personal protective equipment, are also provided in this section. |
| Physical and chemical properties | Specific chemical information, such as boiling points, vapor pressures, and flash points are included in this section. |
| Stability and reactivity | Some chemicals become unstable after a period of time or storage conditions, or may react with other chemicals. Any information regarding this type of information is provided here. |

(continued)

Table 10-2 *(continued)*

Toxicology information	The long-term health effects and other concerns regarding acute and chronic exposures are provided in this section.
Ecological information	Information regarding any potential environmental effects are listed here.
Disposal considerations	Although in many cases this section states "follow local regulations," this is the section that outlines any regulatory requirements for disposal.
Transport information	Information regarding the DOT regulations is listed here, typically the hazard class and UN identification number are provided in this section.
Regulatory information	Any other regulations that apply to the use, storage, or disposal of the chemical are listed here. If the chemical is covered by other regulations, this information is listed in this section.

Chemical Manufacturers Association (CMA)
An association of the manufacturers of hazardous chemicals which also funds CHEMTREC.

CHEMTREC
A national communications center that provides technical advice and guidance for emergency responders at the scene of a chemical emergency.

during an emergency situation. The **Chemical Manufacturers Association (CMA)** has made information available to ensure that the latest data are available to first responders. One method of obtaining this information is via Internet through the **Chemical Transportation Emergency Center** or **CHEMTREC.**

Supplementing the telephone hotline, the Chemical Manufacturers Association announced the creation of an online database system that will be updated daily by the chemical manufacturers. Using the MSDS as the starting point, first responders can access the system via the internet to obtain the most up-to-date and accurate information. The guide to CHEMTREC for emergency responders can be downloaded at http://www.chemtrec.com/Chemtrec/Resources/UserGuid.htm.

Environmental Protection Agency

The primary mission of the EPA is to protect and enhance our environment. In conducting this mission, it assumes the responsibility as the lead agency for carrying out Title III reporting and training requirements. The EPA also has responsibility for hazardous waste site operations and superfund waste site cleanup activities. The EPA also serves as the chair for the fourteen federal agencies that participate as the National Response Team. As discussed, the NRT is available to local communities to provide assistance and coordination between federal, state, and local agencies.

(a)

AIR PRODUCTS

Material Safety Data Sheet
Version 1.2
Revision Date 08/12/2003
MSDS Number 300000000099
Print Date 10/01/2003

1. PRODUCT AND COMPANY IDENTIFICATION

Product name : Nitrogen

Chemical formula : N2

Synonyms : Nitrogen, Nitrogen gas, Gaseous Nitrogen, GAN

Product Use Description : General Industrial

Company : Air Products and Chemicals, Inc
7201 Hamilton Blvd.
Allentown, PA 18195-1501

Telephone : 800-345-3148

Emergency telephone number : 800-523-9374 USA
01-610-481-7711 International

2. COMPOSITION/INFORMATION ON INGREDIENTS

Components	CAS Number	Concentration (Volume)
Nitrogen	7727-37-9	100 %

Concentration is nominal. For the exact product composition, please refer to Air Products technical specifications.

3. HAZARDS IDENTIFICATION

Emergency Overview

High pressure gas.
Can cause rapid suffocation.
Self contained breathing apparatus (SCBA) may be required.

Potential Health Effects

Inhalation : In high concentrations may cause asphyxiation. Asphyxiation may bring about unconsciousness without warning and so rapidly that victim may be unable to protect themselves.

Eye contact : No adverse effect.

Skin contact : No adverse effect.

Ingestion : Ingestion is not considered a potential route of exposure.

Chronic Health Hazard : Not applicable.

Exposure Guidelines

Air Products and Chemicals, Inc   1/7   Nitrogen

(b)

Material Safety Data Sheet
Version 1.2
Revision Date 08/12/2003
MSDS Number 300000000099
Print Date 10/01/2003

Primary Routes of Entry : Inhalation

Target Organs : None known.

Symptoms : Exposure to oxygen deficient atmosphere may cause the following symptoms: Dizziness. Salivation. Nausea. Vomiting. Loss of mobility/consciousness.

Aggravated Medical Condition
None.

Environmental Effects
Not harmful.

4. FIRST AID MEASURES

General advice : Remove victim to uncontaminated area wearing self contained breathing apparatus. Keep victim warm and rested. Call a doctor. Apply artificial respiration if breathing stopped.

Eye contact : Not applicable.

Skin contact : Not applicable.

Ingestion : Ingestion is not considered a potential route of exposure.

Inhalation : Remove to fresh air. If breathing is irregular or stopped, administer artificial respiration. In case of shortness of breath, give oxygen.

5. FIRE-FIGHTING MEASURES

Suitable extinguishing media : All known extinguishing media can be used.

Specific hazards : Upon exposure to intense heat or flame, cylinder will vent rapidly or rupture violently. Product is nonflammable and does not support combustion. Move away from container and cool with water from a protected position. Keep containers and surroundings cool with water spray.

Special protective equipment for fire-fighters : Wear self contained breathing apparatus for fire fighting if necessary.

6. ACCIDENTAL RELEASE MEASURES

Personal precautions : Evacuate personnel to safe areas. Wear self-contained breathing apparatus when entering area unless atmosphere is proved to be safe. Monitor oxygen level. Ventilate the area.

Environmental precautions : Do not discharge into any place where its accumulation could be dangerous. Prevent further leakage or spillage if safe to do so.

Air Products and Chemicals, Inc   2/7   Nitrogen

(c)

Material Safety Data Sheet
Version 1.2
Revision Date 08/12/2003
MSDS Number 300000000099
Print Date 10/01/2003

Methods for cleaning up : Ventilate the area.

Additional advice : If possible, stop flow of product. Increase ventilation to the release area and monitor oxygen level. If leak is from cylinder or cylinder valve, call the Air Products emergency telephone number. If the leak is in the user's system, close the cylinder valve, safely vent the pressure, and purge with an inert gas before attempting repairs.

7. HANDLING AND STORAGE

Handling

Protect cylinders from physical damage; do not drag, roll, slide or drop. Do not allow storage area temperature to exceed 50°C (122°F). Only experienced and properly instructed persons should handle compressed gases. Before using the product, determine its identity by reading the label. Know and understand the properties and hazards of the product before use. When doubt exists as to the correct handling procedure for a particular gas, contact the supplier. Do not remove or deface labels provided by the supplier for the identification of the cylinder contents. When moving cylinders, even for short distances, use a cart (trolley, hand truck, etc.) designed to transport cylinders. Leave valve protection caps in place until the container has been secured against either a wall or bench or placed in a container stand and is ready for use. Use an adjustable strap wrench to remove over-tight or rusted caps. Before connecting the container, check the complete gas system for suitability, particularly for pressure rating and materials. Before connecting the container for use, ensure that back feed from the system into the container is prevented. Ensure the complete gas system is compatible for pressure rating and materials of construction. Ensure the complete gas system has been checked for leaks before use. Employ suitable pressure regulating devices on all containers when the gas is being emitted to systems with lower pressure rating than that of the container. Never insert an object (e.g. wrench, screwdriver, pry bar, etc.) into valve cap openings. Doing so may damage valve, causing a leak to occur. Open valve slowly. If user experiences any difficulty operating cylinder valve discontinue use and contact supplier. Close container valve after each use and when empty, even if still connected to equipment. Never attempt to repair or modify container valves or safety relief devices. Damaged valves should be reported immediately to the supplier. Close valve after each use and when empty. Replace outlet caps or plugs and container caps as soon as container is disconnected from equipment. Do not subject containers to abnormal mechanical shocks which may cause damage to their valve or safety devices. Never attempt to lift a cylinder by its valve protection cap or guard. Do not use containers as rollers or supports or for any other purpose than to contain the gas as supplied. Never strike an arc on a compressed gas cylinder or make a cylinder a part of an electrical circuit. Do not smoke while handling product or cylinders. Never re-compress a gas or a gas mixture without first consulting the supplier. Never attempt to transfer gases from one cylinder/container to another. Always use backflow protective device in piping. When returning cylinder install valve outlet cap or plug leak tight. Never use direct flame or electrical heating devices to raise the pressure of a container. Containers should not be subjected to temperatures above 50°C (122°F). Prolonged periods of cold temperature below -30°C (-20°F) should be avoided.

Storage

Full containers should be stored so that oldest stock is used first. Containers should be stored in a purpose build compound which should be well ventilated, preferably in the open air. Stored containers should be periodically checked for general condition and leakage. Observe all regulations and local requirements regarding storage of containers. Protect containers stored in the open against rusting and extremes of weather. Containers should not be stored in conditions likely to encourage corrosion. Containers should be stored in the vertical position and properly secured to prevent toppling. The container valves should be tightly closed and where appropriate valve outlets should be capped or plugged. Container valve guards or caps should be in place. Keep containers tightly closed in a cool, well-ventilated place. Store containers in location free from fire risk and away from sources of heat and ignition. Full and empty cylinders should be segregated. Do not allow storage temperature to exceed 50°C (122°F). Return empty containers in a timely manner.

Air Products and Chemicals, Inc   3/7   Nitrogen

Figure 10-4 *Sample MSDS.* Source: Firefighter's Handbook 3rd ed. pg 945–946.

Figure 10-5 *Marine pollutant indicator.*

National Oceanic and Atmospheric Administration (NOAA)
An organization that works with the EPA to develop a chemical reactivity worksheet that may be useful to first responders.

National Institute of Occupational Safety and Health (NIOSH)
An institute that provides toxicity testing and standards and publishes the *Guide to Chemical Hazards.*

Federal Bureau of Investigation (FBI)
Under the direction of the Department of Justice, this federal agency is responsible for the criminal investigation of acts of terrorism.

National Domestic Preparedness Office (NDPO)
The coordination center for all federal efforts in response to weapons of mass destruction.

National Oceanic and Atmospheric Administration and the EPA

The U.S. Department of Commerce, the **National Oceanic and Atmospheric Administration (NOAA),** and the EPA together have developed a chemical reactivity worksheet for first responders. The electronic worksheet provides the hazardous properties of chemicals and gives the hazards involved with mixing a number of chemicals as specified by the user. More about the NOAA can be found at http://www.etl.noaa.gov/.

U.S. Coast Guard

The Coast Guard offers a chemical and hazardous response information system (COMDTINST 16465.12C). This CD-ROM information base contains physical data and emergency procedures in a searchable format. Local fire departments have the responsibility of notifying the Coast Guard when a hazardous materials release will impact the coastal waters. The EPA has the responsibility for inland waters. Figure 10-5 shows the approved warning label for materials that pollute water. More information is available on the Coast Guard's website, http://www.uscg.mil/vrp/.

National Institute of Occupational Safety and Health

The **National Institute of Occupational Safety and Health (NIOSH)** is a federal agency under the U.S. Department of Health and Human Services that is responsible for investigating the toxicity of workroom environments and all other matters relating to safe industrial practices. NIOSH publishes the *Guide to Chemical Hazards,* an excellent source for obtaining the health hazards relating to hazardous materials. It can be obtained at http://www.cdc.gov/niosh/npg/. Other activities of NIOSH are the testing and certification of respiratory and protective devices along with the testing of air sampling detector tubes. Hazardous materials experts have recommended that NIOSH work with the U.S. military to provide protective breathing and clothing standards for protection from weapons of mass destruction (U.S. Department of Health and Human Services [HHS], 2000).

Federal Bureau of Investigation and the National Domestic Preparedness Office

Weapons of mass destruction (WMD)
The term contains the use of hazardous materials, nuclear radiation, biological agents, chemicals, explosives, and cyber and agriculture agents.

Under the leadership of the **Federal Bureau of Investigation (FBI)**, the **National Domestic Preparedness Office (NDPO)** is the clearinghouse for all **weapons of mass destruction (WMD).** The NDPO serves as a single program and policy co-ordination office for domestic preparedness programs for state and local communities. The NDPO is created to act only as a coordinating and facilitating point for all efforts in training, information dissemination, and standards and policy development relating to terrorist incidence response (HHS, 2000).

Department of Defense

Department of Defense (DOD)
A federal department that supplies manpower, training, and equipment for a full range of hazardous substances which may be encountered.

Several sources in the **Department of Defense (DOD)** have interests and expertise in biological and/or chemical terrorism. Currently, the U.S. Army Soldier and Biological Command (SBCCOM) at the Aberdeen proving ground administers the domestic preparedness training program. This program is designed to assist in the training of first responders in the largest U.S. cities.

The SBCCOM has the responsibility for developing the standards and tests for chemical protective clothing, which in many cases is far superior to the protective clothing now available in the civilian marketplace. Although some of the information is classified, specific assistance may be provided, if requested, through the FBI/NDPO representative. SBCCOM also maintains a hotline providing real-time referrals that can be utilized by local agencies. They can be contacted at the Domestic Preparedness help line, 800-368-6498 (HHS, 2000).

Department of Energy and the Nuclear Regulatory Commission

Department of Energy (DOE)
The federal department that has primary responsibility involving radioactive waste generated by the nuclear weapons program or by the nuclear reactors supplying energy.

In the past, the **Department of Energy (DOE)** and the **Nuclear Regulatory Commission (NRC)** have provided training to state and local emergency personnel, who would assist in a response to an incident/emergency at a DOE or an NRC licensed facility. More recently, the DOE has started to provide training to state and local emergency personnel along major U.S. transportation corridors that carry radiological materials. Further information on the responsibilities and services provided by the DOE and the NRC can be obtained at http://www.nrc.gov/about-nrc/emerg-preparedness.html.

Nuclear Regulatory Commission (NRC)
A commission within the Department of Energy, which recommends and establishes controls by providing licenses to facilities that use nuclear products.

Nuclear Radiation Releases Since the 1950s, use of nuclear materials has continued to increase worldwide. In addition to nuclear fuels used for the production of electric power, nuclear materials are widely used for medical purposes, industrial controls, and household uses (e.g., ionization smoke detectors) as well as for basic research and weapons military systems. The global use of nuclear materials continues to expand as well as the opportunities for accidental release as more radioactive materials are being transported and handled.

The threat of an intentional release of ionizing radiation materials is now seen as a possibility because of the significant increase in terrorism. As firefighters across the nation are the first responders, there is a need for training to handle radiological incidents, and for protection against direct exposure to radioactive materials and their radiated waves and resulting contamination. See previous discussion on DOE and NRC training programs.

Radioactive Materials Radioactive materials are numerous and present themselves in a number of different forms. They can be the emissions from a damaged energy generating plant, the result of a release from a nuclear weapon, or an accident in a medical or research laboratory. They too can be used by terrorists as a threat for concessions to demands or to be released to cause death and injuries.

There are materials which undergo spontaneous transformation and release radiant energy or atomic particles. Three forms of radioactivity that are of a concern to firefighters are **alpha radiation, beta radiation,** and **gamma radiation.** Alpha radiation particles are the least dangerous as long as the particles remain outside the body. Protective clothing and respiratory protection can provide protection against alpha radiation. Beta radiation particles have a higher energy level than alpha radiation and can travel through the air from 10 to 100 ft (3 to 30 m). Turnout clothing and respiratory protection generally provide enough protection against beta particles, but they are hazardous if swallowed or if they contact the skin.

The most dangerous form of ionizing radiation is gamma radiation. It consists of electromagnetic rays similar to x-rays, and these rays easily penetrate the human body. Protection from external radiation can be accomplished three ways: time, distance, and shielding.

The shorter amount of time exposure to radiation the smaller the radiation dose, so first responders must work to reduce the amount of time exposed. One method might be to rotate the assignment to limit the amount of time exposure by an individual.

The further the distance from the source of radiation, the smaller the dose. As the distance doubles, the amount of radiation decreases in proportion to the square of the distance. In other words, doubling the distance from the source reduces the radiation exposure by one-fourth.

Certain dense materials such as lead, concrete, or earth and water will stop some of the radiation rays. The thickness of the shielding needed depends on the type of the material giving off the radiation, the amount of radiation, and the distance from the source. Figure 10-6 indicates the stopping power of various materials.

In all cases involving ionizing radiation, the wearing of full protective clothing and SCBA is required. However, caution must be taken because some situations require additional protective equipment to enter the area safely. Figure 10-7 shows an example of radiation protective clothing for firefighters.

Radioactive Materials Advantages and Disadvantages

Almost any radioactive material can be used to construct a **radiological dispersal device (RDD)** such as fission products, spent fuel from nuclear reactors, and

---

Alpha radiation
Radiation that is given off by alpha particles.

Beta radiation
Radiation that is given off by beta particles, which are smaller than alpha particles but have more penetrating power.

Gamma radiation
The most dangerous form of penetrating radiation. Gamma rays have 10,000 times the penetrating power of alpha particles and 100 times the penetrating power of beta particles.

Radiological dispersal device (RDD)
A device such as an explosive used to disperse radiological products such as fission products which are the waste from fuel of nuclear reactors or medical, industrial, and research radiological waste.

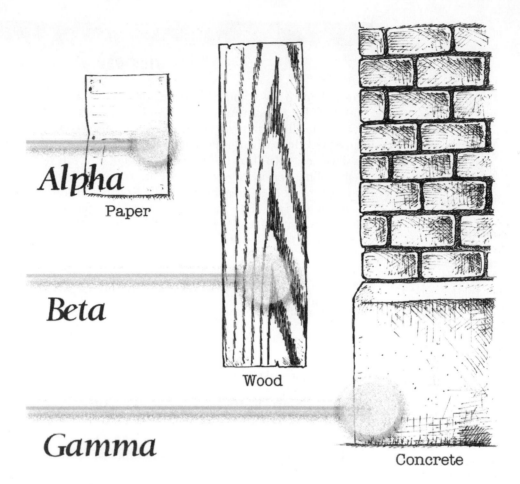

Figure 10-6
Radiation penetration.

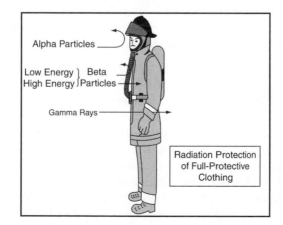

Figure 10-7
Radiation protective clothing for firefighters.

Rad
A measure of the dose of ionizing radiation absorbed by body tissues stated in terms of the amount of energy absorbed per unit of mass of tissue.

Relative biological effectiveness (RBE)
A measure of how damaging a given type of particle is when compared to an equivalent dose of x-rays.

Rem
A measure of the dose of ionizing radiation to body tissue stated in terms of its estimated biological effect relative to a dose of one roentgen (r) of x-rays.

Measuring Radiation

We measure absorbed radiation doses with a unit termed a **rad.** A rad is defined by the amount of energy deposited in the tissue of the person receiving the radiation. To more accurately assess the risk of radiation, the absorbed dose of energy in rads is multiplied by the **relative biological effectiveness (RBE)** of the radiation to get the biological dose equivalent in **rems.** The acronym *rem* stands for roentgen equivalent man. This is the important dosage measurement for first responders as there is a maximum amount of rem that first responders can be exposed to without serious health consequences.

medical, industrial, and research waste. For the greatest damaging effect, the most radioactive of fissile material would be used as the active ingredient in the RDD.

An RDD would likely have a significant impact on the target population. An uninformed society may contribute to the terrorist's objectives because they are more likely to initiate widespread panic and disperse away from the target area. Unless the public has a basic understanding of nuclear biological chemical (NBC) terrorism, any use of RDD is likely to invoke the fear of nuclear war.

The disadvantage of RDDs over nuclear weapons involves the use of explosive materials to spread large amounts of radioactive materials greater distances. If using radioactive waste, which is different than fissile material, the terrorists and the first responder handlers are more likely to become inadvertently contaminated by the radiation, since radioactive wastes are highly contaminable. Of importance for first responders is to determine the possibility of safely entering the contaminated areas by having the capability to measure the radiation present (see the above textbox) and protect themselves while entering the area.

Radiation Exposures to First Responders The amount of emergency exposure to radiation by first responders is usually allowed to exceed those amounts considered tolerable to persons who work continuously with radioactive materials. This is accomplished by raising the exposure, within limits for single doses. This is considered acceptable as long as certain limits are not exceeded. The National Council on Radiation Protection and Measurement has recommended that in a life-saving action, such as a search for and removal of injured persons or for entry to prevent conditions that would injure or kill numerous persons, entry may be permitted as long as the whole body exposure does not exceed 100 rem.

TERRORISM

In addition to hazardous materials, firefighters are faced with other materials that present a serious safety threat. Since the early 1960s, many terrorist acts have taken place in Northern Ireland and the Middle East. Most Americans, although concerned about the loss of innocent lives, were not personally impacted as these acts occurred overseas. Starting in the late 1970s and into the 1980s, terrorist acts against Americans stationed overseas occurred more often, and within U.S. borders more acts of violence began to occur. The situation in the United States has escalated since the early 1990s. The 1993 bombing of the World Trade Center, the 1995 bombing of the Alfred P. Murrah Federal Building in Oklahoma City, and the subsequent 2001 World Trade Center air attack all point out that first responders, as well as the American public, face a serious threat.

During the same period, terrorists began to use a variety of materials to attack their targets. Today we term these materials used by terrorists weapons of mass destruction (WMD). The term contains the use of hazardous materials, the use of nuclear radiation discussed previously, as well as biological agents, chemicals, explosives, and cyber and agriculture agents. All are designed to kill or seriously injure numerous people. First responders must to be trained, equipped, and prepared to handle these attacks efficiently and safely.

WMD attacks are more difficult and different than the usual incident response. The four areas of difficulties are:

1. A large number of persons needing immediate assistance at the same time;
2. Multifunctions, some of a technical nature, conducted simultaneously;
3. The immediate involvement of federal and state agencies; and
4. The overwhelming response of the national media.

At the World Trade Center attack in 2001, numerous persons needed urgent and immediate assistance from all emergency responding agencies at the same time. This incident pointed out the need and importance of carrying out multifunctions. The fire service was impacted with the need for both evacuation and fire attack to save trapped persons. The emergency medical service was very active with numerous persons injured and having to make arrangements for the expected deluge of injured persons at the hospitals. Law enforcement personnel were busy as well, controlling traffic, providing crowd control, and assisting in evacuation. This incident required multifunctions to be carried out simultaneously and be coordinated in a manner to optimize limited resources.

Department of Homeland Security (DHS)
A federal department responsible for protecting U.S. territory from terrorist attacks and responding to natural disasters.

Soon, federal governmental agencies arrived at the scene. The FBI wanted to start the crime investigation process while the incident was still active. This need to conduct multiple functions provided by local, state, and federal agencies at the same time and in the same location while still maintaining overall coordination and control resulted in the 2003 presidential directive 5 (HSPD-5) to the **Department of Homeland Security (DHS)** to establish and develop a National Incident Management System (NIMS).

HSPD-5 was followed by directives HSPD-7 and HSPD-8, which required NIMS to be "not just a new version of Incident Command System (ICS) but to incorporate all aspects of preparing for and managing incidents." This new comprehensive system contains the following components.

- Command and management
- Preparedness resource management
- Communications and information management
- Supporting technologies and ongoing management and maintenance

ICS is the command and management component. The NIMS system will be used to provide the coordination and control of multijurisdictional incidents. Joint multifunctional efforts need to have a comprehensive and coordinated plan which will be developed using the NIMS under the direction of a **unified command (UC).**

Resulting from the circumstances of 9/11, the news media on a national scale immediately overwhelmed the incident, needing answers that were not yet available. At a WMD incident such as this, the information flowing from the news media may have a national impact on the public confidence in government; therefore, news releases must be carefully constructed, accurate, and complete.

All of these factors make a WMD incident a challenging, difficult task that can only be successfully handled with careful preincident planning, joint exercises, and training to use a variety of specialized equipment.

Unified command (UC)
The structure used to manage emergency incidents involving multiple jurisdictions or multiple response agencies that have responsibility for control of the incident.

SUMMARY

Since the 1940s, there has been a dramatic increase in the number and variety of hazardous materials, resulting in more hazardous materials incidents. As the amount of transportation and handling of these materials grows daily, so too do the incidents. The increase in responding to unwanted releases of these substances has prompted federal and state governments to put into effect regulations to manage the use and control of these substances.

First responders in the United States have developed warning systems to assist them in the identification of hazardous materials. These systems are the DOT and the NFPA placard systems. The DOT identifies materials being transported across state lines and the NFPA system identifies buildings where the materials are handled or stored.

Starting with the original 1993 World Trade Center bombing, followed by the 1995 Oklahoma City federal building bombing, and then the 2001 terrorist attacks on the World Trade Center, there has been a rapid increase in the use of substances that we now define as weapons of mass destruction (WMD). In response to this increase in terrorist activities, the National Incident Management System has been developed to plan for and coordinate the efforts of all agencies concerned with incidents of WMD.

One way fire departments can prepare for these types of incidents is by reviewing lessons learned by other first responders and building upon these learning experiences. Using the information gained from the experiences of others will assist them to safely control and protect citizens and communities.

KEY TERMS

Alpha radiation Radiation that is given off by alpha particles. They are the least dangerous of the three radiation components that may be encountered by firefighters. Alpha is the least dangerous as an external radiation hazard and the most dangerous as internal radiation. Alpha particles have very low penetrating power and as a result seldom penetrate the skin. Turnout clothing will provide the needed protection from alpha radiation.

Beta radiation Radiation that is given off by beta particles, which are smaller than alpha particles but have more penetrating power. Large amounts of beta radiation externally can seriously damage skin tissue. Beta particles enter the body through damaged skin by ingestion or inhalation and will cause organ damage similar to that caused by alpha particles. Full protective clothing including SCBA is required to protect against beta radiation.

Chemical Manufacturers Association An association of the manufacturers of hazardous chemicals which also funds CHEMTREC.

CHEMTREC A national communications center that provides technical advice and guidance for emergency responders at the scene of a chemical emergency. This network of chemical manufacturers provides emergency information and response teams if necessary.

Department of Defense (DOD) A federal department that supplies manpower, training, and equipment for a full range of hazardous substances which may be encountered. DOD laboratories and bases can be a source of expertise, equipment, and supplies for use in local emergencies.

Department of Energy (DOE) The federal department that has primary responsibility involving radioactive waste generated by the nuclear weapons program or by the nuclear reactors supplying energy.

Department of Homeland Security (DHS) A federal department responsible for protecting U.S. territory from terrorist attacks and responding to natural disasters.

Department of Transportation (DOT) A federal department responsible for issuing standards and regulations relating to the transportation of hazardous materials nationwide. The DOT also trains and inspects carriers and shippers of hazardous materials to ensure that they are in full compliance.

Emergency Planning and Community Right-to-Know Act (EPCRA) An act that encourages and supports emergency planning efforts at the state and local levels and provides the public and local governments with information concerning potential chemical hazards present in their communities.

***Emergency Response Guidebook* (ERG)** A booklet that contains the chemical names and the commonly used names for most hazardous chemicals. It also has an identification numbering system as well as a placard system for transportation vehicles and containers.

Federal Bureau of Investigation (FBI) Under the direction of Homeland Security, this federal agency is responsible for the criminal investigation of acts of terrorism.

Gamma radiation The most dangerous form of penetrating radiation. Gamma rays have 10,000 times the penetrating power of alpha particles and 100 times the penetrating power of beta particles. They pass through the body and strike body organs, and in doing so cause extensive internal damage.

Hazardous materials Any material in any form or quantity that poses an unreasonable risk to safety and health and property when transported in commerce (DOT 49 CFR 171).

Hazardous substance Any substance designated under the Clean Water Act and the

Comprehensive Environmental Response Compensation and Liability Act (CERCLA) as posing a threat to waterways and the environment when released. When used by OSHA, hazardous substances refers to any substance defined by EPA, within section 101 of CERCLA; or as any biological agent or other disease-causing agent as defined by EPA within section 101 of CERCLA; or as any substance listed by DOT as a hazardous material and any hazardous waste as defined by EPA in 40 CFR 261.3 or by DOT in 49 DFR 171.8.

Local emergency planning committee (LEPC) A committee that is responsible for developing and maintaining a local emergency response plan that ensures quick and effective response to chemical emergencies. The plan provides effective actions to be taken and methods of evacuating if needed.

Materials Safety Data Sheets (MSDS) A required worksheet prepared by manufacturers and marketers to provide workers with detailed information about the properties of a commercial product that contains hazardous substances.

National Domestic Preparedness Office (NDPO) The coordination center for all federal efforts in response to weapons of mass destruction.

National Fire Protection Association (NFPA) 704 system The procedure of assigning one of the five numbers, 0 through 4, to the relative degree of three hazards, which are color coded (health, flammability, and chemical reactivity) for a given hazardous substance. A diamond-shaped figure is divided into four quadrants, each of which is color coded for each of three hazards (blue for health hazards, red for flammability, and yellow for chemical reactivity) and numbered 0 to 4 to designate the degree of the relevant hazard. The number 0 indicates that the materials are considered to be stable even in the event of fire.

National Institute of Occupational Safety and Health (NIOSH) An institute that provides toxicity testing and standards, publishes the *Guide to Chemical Hazards,* an important source of information for first responders, and is working to develop protective breathing and clothing standards for the weapons of mass destruction.

National Oceanic and Atmospheric Administration (NOAA) An organization that works with the EPA to develop a chemical reactivity worksheet that may be useful to first responders.

National Response Team (NRT) A team of fourteen federal agencies assembled to assist state and local agencies with hazardous materials incidents and the preplanning for these incidents.

Nuclear Regulatory Commission (NRC) A commission within the Department of Energy, which recommends and establishes controls by providing licenses to facilities that use nuclear products.

Rad A measure of the dose of ionizing radiation absorbed by body tissues stated in terms of the amount of energy absorbed per unit of mass of tissue.

Radiological dispersal device (RDD) A device such as an explosive used to disperse radiological products such as fission products which are the waste from fuel of nuclear reactors or medical, industrial, and research radiological waste.

Relative biological effectiveness (RBE) A measure of how damaging a given type of particle is when compared to an equivalent dose of x-rays.

Rem A measure of the dose of ionizing radiation to body tissue stated in terms of its estimated biological effect relative to a dose of one roentgen (r) of x-rays.

State emergency response committee (SERC) Appointed by the governor, this committee is responsible for the coordination of emergency plans, which are developed by the LEPCs.

Unified command (UC) The structure used to manage emergency incidents involving multiple jurisdictions or multiple response agencies that have responsibility for control of the incident.

Weapons of mass destruction (WMD) The term contains the use of hazardous materials, nuclear radiation, biological agents, chemicals, explosives, and cyber and agriculture agents. All are designed to kill or seriously injure numerous people.

REVIEW QUESTIONS

1. Describe the two types of hazardous materials warning systems and their benefits to first responders.

2. List the three types of radiation and give examples of how to protect responders from each type.

3. Describe the potential benefits of the LEPC for first responders.

4. Explain what an RDD is and how it will affect a response.

REFERENCES

Angle, James S. (2005). *Occupational Safety and Health in the Emergency Services* (2d ed.) (Clifton Park, NY: Delmar Cengage Learning).

Environmental Protection Agency Counterterrorism Unit, http://www.epa.gov/ebtpages/emercounter-terrorism.html.

Fitch, J. Patrick, Ellen Raber and Dennis R. Imbro. (2003). "Technology Challenges in Responding to Biological or Chemical Attacks in the Civilian Sector," *Science* 302(5649), 1350–1354.

Gordon, John C. and Steen Bech-Nielsen. (1999). "Biological Terrorism: A Direct Threat to Our Livestock Industry," *Military Medicine*.

Hawley, Chris. (2008). *Hazardous Materials Incidents* (3d ed.) (Clifton Park, NY: Delmar Cengage Learning).

International Association of Fire Chiefs. (2005). "IAFC Survey," *Fire Engineering* 158(1). http://www.fireengineering.com/display_article/222905/25/none/none/INCID/Haz-Mat-Placards-Critical-To-Emergency-Response,-Survey-Show

Mangold, Tom and Jeff Goldberg. (2000). *Plague Wars* (New York: St. Martin's Press).

Nuclear, Biological & Chemical Warfare, http://www.nbc-links.com/index1.html.

Randall, Stephen. (1995). "Chemical Manufacturers Association Creates MSDS Data Base," *Fire Engineering* 148(6).

Superfund Benefits Analysis, http://www.epa.gov/superfund/accomp/news/benefits.htm. This comprehensive report outlines the legislation, funding, and benefits to first responders as well as to local communities.

U.S. Department of Health and Human Services. (2000, February). "Chemical, Biological and Respiratory Protection," workshop sponsored by NIOSH-DOD-OSHA.

U.S. Department of Labor and Occupational Safety and Health Administration. Safety and health topics, http://www.osha.gov/SLTC/radiation/index.html and http://www.osha.gov/SLTC/biologicalagents/index.html.

U.S. Department of Transportation. (2008). *Emergency Response Guidebook* (Washington, DC: Government Printing Office). Copies are provided by the DOT to first responders at no cost. A copy can be downloaded at the EPA website, http://www.epa.gov/epahome/abouepa.htm.

Walsh, Donald et al. (2005). *National Incident Management System. Principle and Practices* (Sudbury, MA: Jones and Bartlett Publishers).

Appendix

A

REFERENCE TABLES

Inches converted to decimals of a foot.

Inches		Decimal of a Foot	Inches		Decimal of a Foot
0	1/8	.010416	6	1/8	.510416
	1/4	.020833	(.50)	1/4	.520833
	3/8	.031250		3/8	.531250
	1/2	.041666		1/2	.541666
	5/8	.052083		5/8	.552083
	3/4	.062500		3/4	.562500
	7/8	.072916		7/8	.572916
1	1/8	.093750	7	1/8	.593750
(.083333)	1/4	.104166	(.583333)	1/4	.604166
	3/8	.114583		3/8	.614583
	1/2	.125000		1/2	.625000
	5/8	.135416		5/8	.635416
	3/4	.145833		3/4	.645833
	7/8	.156250		7/8	.656250
2	1/8	.177083	8	1/8	.677083
(.166666)	1/4	.187500	(.666666)	1/4	.687500
	3/8	.197916		3/8	.697916
	1/2	.208333		1/2	.708333
	5/8	.218750		5/8	.718750
	3/4	.229166		3/4	.729166
	7/8	.239583		7/8	.739583
3	1/8	.260416	9	1/8	.760416
(.250)	1/4	.270833	(.750)	1/4	.770833
	3/8	.281250		3/8	.781250
	1/2	.291666		1/2	.791666
	5/8	.302083		5/8	.802083
	3/4	.312500		3/4	.812500
	7/8	.322916		7/8	.822916
4	1/8	.343750	10	1/8	.843750
(.333333)	1/4	.354166	(.833333)	1/4	.854166
	3/8	.364583		3/8	.864583
	1/2	.375000		1/2	.875000
	5/8	.385416		5/8	.885416
	3/4	.395833		3/4	.895833
	7/8	.406250		7/8	.906250
5	1/8	.427083	11	1/8	.927083
(.416666)	1/4	.437500	(.916666)	1/4	.937500
	3/8	.447916		3/8	.947916
	1/2	.458333		1/2	.958333
	5/8	.468750		5/8	.968750
	3/4	.479166		3/4	.979166
	7/8	.489583		7/8	.989583

Decimal equivalents of fractions of an inch.

Inches	Decimal of an Inch	Inches	Decimal of an Inch
1/64	.015625	33/64	.515625
1/32	.03125	17/32	.53125
3/64	.046875	35/64	.546875
1/16	.0625	9/16	.5625
5/64	.078125	37/64	.578125
3/32	.09375	19/32	.59375
7/64	.109375	39/64	.609375
1/8	.125	5/8	.625
9/64	.140625	41/64	.640625
5/32	.15625	21/32	.65625
11/64	.171875	43/64	.671875
3/16	.1875	11/16	.6875
13/64	.203125	45/64	.703125
7/32	.21875	23/32	.71875
15/64	.234375	47/64	.734375
1/4	.25	3/4	.75
17/64	.265625	49/64	.765625
9/32	.28125	25/32	.78125
19/64	.296875	51/64	.796875
5/16	.3125	13/16	.8125
21/64	.328125	53/64	.828125
1/3	.333	27/32	.84375
11/32	.34375	55/64	.859375
23/64	.359375	7/8	.875
3/8	.375	57/64	.890625
25/64	.390625	29/32	.90625
13/32	.40625	59/64	.921875
27/64	.421875	15/16	.9375
7/16	.4375	61/64	.953125
29/64	.453125	31/32	.96875
15/32	.46875	63/64	.984375
31/64	.484375	1	1.
1/2	.5		

Minutes converted to decimals of a degree.

Min.	Deg.	Min.	Deg.	Min.	Deg.	Min.	Deg.	Min.	Deg.	Min.	Deg.
1	.0166	11	.1833	21	.3500	31	.5166	41	.6833	51	.8500
2	.0333	12	.2000	22	.3666	32	.5333	42	.7000	52	.8666
3	.0500	13	.2166	23	.3833	33	.5500	43	.7166	53	.8833
4	.0666	14	.2333	24	.4000	34	.5666	44	.7333	54	.9000
5	.0833	15	.2500	25	.4166	35	.5833	45	.7500	55	.9166
6	.1000	16	.2666	26	.4333	36	.6000	46	.7666	56	.9333
7	.1166	17	.2833	27	.4500	37	.6166	47	.7833	57	.9500
8	.1333	18	.3000	28	.4666	38	.6333	48	.8000	58	.9666
9	.1500	19	.3166	29	.4833	39	.6500	49	.8166	59	.9833
10	.1666	20	.3333	30	.5000	40	.6666	50	.8333	60	1.0000

Standard conversions.

To Change	To	Multiply By	To Change	To	Multiply By
Inches	Feet	0.0833	Inches of mercury	Inches of water	13.6
Inches	Millimeters	25.4	Inches of mercury	Feet of water	1.1333
Feet	Inches	12	Inches of mercury	Pounds per square inch	0.4914
Feet	Yards	0.3333	Ounces per square inch	Inches of mercury	0.127
Yards	Feet	3	Ounces per square inch	Inches of water	1.733
Square inches	Square feet	0.00694	Pounds per square inch	Inches of water	27.72
Square feet	Square inches	144	Pounds per square inch	Feet of water	2.310
Square feet	Square yards	0.11111	Pounds per square inch	Inches of mercury	2.04
Square yards	Square feet	9	Pounds per square inch	Atmospheres	0.0681
Cubic inches	Cubic feet	0.00058	Feet of water	Pounds per square inch	0.434
Cubic feet	Cubic inches	1728	Feet of water	Pounds per square foot	62.5
Cubic feet	Cubic yards	0.03703	Feet of water	Inches of mercury	0.8824
Cubic yards	Cubic feet	27	Atmospheres	Pounds per square inch	14.696
Cubic inches	Gallons	0.00433	Atmospheres	Inches of mercury	29.92
Cubic feet	Gallons	7.48	Atmospheres	Feet of water	34
Gallons	Cubic inches	231	Long tons	Pounds	2240
Gallons	Cubic feet	0.1337	Short tons	Pounds	2000
Gallons	Pounds of water	8.33	Short tons	Long tons	0.89285
Pounds of water	Gallons	0.12004			
Ounces	Pounds	0.0625			
Pounds	Ounces	16			
Inches of water	Pounds per square inch	0.0361			
Inches of water	Inches of mercury	0.0735			
Inches of water	Ounces per square inch	0.578			
Inches of water	Pounds per square foot	5.2			

Conversion factors for water.

1 U.S. gallon = 8.3356 pounds
1 U.S. gallon = 0.1337 cubic feet
1 U.S. gallon = 231 cubic inches
1 U.S. gallon = 0.83356 Imperial gallons
1 U.S. gallon = 3.7854 liters

1 Imperial gallon = 10.00 pounds
1 Imperial gallon = 0.16037 cubic feet
1 Imperial gallon = 277.12 cubic inches
1 Imperial gallon = 1.1997 U.S. gallons
1 Imperial gallon = 4.5413 liters

1 liter = 2.202 pounds
1 liter = 0.0353 cubic feet
1 liter = 61.023 cubic inches
1 liter = 0.2642 U.S. gallons
1 liter = 0.2202 Imperial gallons

1 cubic foot of water = 62.355 pounds
1 cubic foot of water = 1728.00 cubic inches
1 cubic foot of water = 7.4805 U.S. gallons
1 cubic foot of water = 6.2355 Imperial gallons
1 cubic foot of water = 28.317 liters

1 pound of water = .01604 cubic feet
1 pound of water = 27.712 cubic inches
1 pound of water = 0.11997 U.S. gallons
1 pound of water = 0.100 Imperial gallons
1 pound of water = 0.45413 liters

1 cubic inch of water = 0.0361 pounds
1 gallon of water = 8.33 pounds

1 inch of water = 0.0361 pounds per square inch
1 foot of water = 0.4334 pounds per square inch
1 pound per square inch = 2.310 feet of water
1 pound per square inch = 2.04 inches of mercury
1 atmosphere = 14.696 pounds per square inch

Geometric formula for area (A), volume (V), and circumference (C), radius (R), and diameter (D).

Circle

$A = 3.142 \times R^2$

$C = 3.142 \times D$

$R = \dfrac{D}{2}$

$D = 2 \times R$

Elipse

$A = 3.142 \times A \times B$

$C = 6.283 \times \dfrac{\sqrt{A^2 + B^2}}{2}$

Parallelogram

$A = H \times L$

Rectangle

$A = W \times L$

Sector of circle

$A = \dfrac{3.142 \times R^2 \times \alpha}{360}$

$L = .01745 \times R \times \alpha$

$\alpha = \dfrac{L}{.01745 \times R}$

$R = \dfrac{L}{.01745 \times \alpha}$

Trapezoid

$A = H \times \dfrac{L_1 + L_2}{2}$

Triangle

$A = \dfrac{W \times H}{2}$

$C^2 = H^2 + \left(\dfrac{W}{2}\right)^2$

Cone

$A_1 = 3.142 \times R \times S + 3.142 \times R^2$

$V = 1.047 \times R^2 \times H$

Cylinder

Area of curved surface =

$(2) \times (3.142) \times (R) \times (H)$

$V = 3.142 \times R^2 \times H$

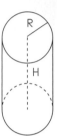

Elliptical tank

$V = 3.142 \times A \times B \times H$

$A_1 = 6.283 \times \dfrac{\sqrt{A^2 + B^2}}{2} \times H + 6.283 \times A \times B$

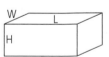
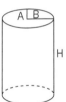

Rectangular solid

$A_1 = 2[W \times L + L \times H + H \times W]$

$V = W \times L \times H$

Sphere

$A_1 = 12.56 \times R^2$

$V = 4.188 \times R^3$

For above figures when used as water containers:

Capacity in gallons $= \dfrac{V}{231}$ when V is in cubic feet

Capacity in gallons $= 7.48 \times V$ when V is in cubic feet

Sizes and capacity of steel pipe.

Nominal Pipe Size (inches)	Outside Diameter (inches)	Inside Diameter (inches)	Capacity of a One Foot Length of Pipe		Nominal Pipe Size (inches)	Outside Diameter (inches)	Inside Diameter (inches)	Capacity of a One Foot Length of Pipe	
			Cubic Feet	Gallons				Cubic Feet	Gallons
Schedule 40 Pipe					Schedule 80 Pipe				
½	0.840	0.622	0.0021	0.0158	½	0.840	0.546	0.0016	0.012
¾	1.050	0.824	0.0037	0.0276	¾	1.050	0.742	0.0030	0.022
1	1.315	1.049	0.0060	0.0449	1	1.315	0.957	0.0050	0.037
1¼	1.660	1.380	0.0104	0.0774	1¼	1.660	1.278	0.0089	0.066
1½	1.900	1.610	0.0142	0.106	1½	1.900	1.500	0.0123	0.092
2	2.375	2.067	0.0233	0.174	2	2.375	1.939	0.0205	0.153
2½	2.875	2.469	0.0332	0.248	2½	2.875	2.323	0.0294	0.220
3	3.500	3.068	0.0513	0.383	3	3.500	2.900	0.0548	0.344
3½	4.000	3.548	0.0686	0.513	3½	4.000	3.364	0.0617	0.458
4	4.500	4.026	0.0883	0.660	4	4.500	3.826	0.0798	0.597
5	5.563	5.047	0.139	1.04	5	5.563	4.813	0.126	0.947
6	6.625	6.065	0.200	1.50	6	6.625	5.761	0.181	1.35
8	8.625	7.981	0.3474	2.60	8	8.625	7.625	0.3171	2.38
10	10.75	10.020	0.5475	4.10	10	10.75	9.564	0.4989	3.74
12	12.75	11.938	0.7773	5.82	12	12.75	11.376	0.7058	5.28
14	14.0	13.126	0.9397	7.03	14	14.0	12.500	0.8522	6.38
16	16.0	15.000	1.2272	9.16	16	16.0	14.314	1.1175	8.36
18	18.0	16.876	1.5533	11.61	18	18.0	16.126	1.4183	10.61
20	20.0	18.814	1.9305	14.44	20	20.0	17.938	1.7550	13.13
24	24.0	22.626	2.7920	20.87	24	24.0	21.564	2.536	19.0

Conversion of inches to millimeters.

in.	mm	in.	mm	in.	mm	in.	mm
1	25.4	26	660.4	51	1295.4	76	1930.4
2	50.8	27	685.8	52	1320.8	77	1955.8
3	76.2	28	711.2	53	1346.2	78	1981.2
4	101.6	29	736.6	54	1371.6	79	2006.6
5	127.0	30	762.0	55	1397.0	80	2032.0
6	152.4	31	787.4	56	1422.4	81	2057.4
7	177.8	32	812.8	57	1447.8	82	2082.8
8	203.2	33	838.2	58	1473.2	83	2108.2
9	228.6	34	863.6	59	1498.6	84	2133.6
10	254.0	35	889.0	60	1524.0	85	2159.0
11	279.4	36	914.4	61	1549.4	86	2184.4
12	304.8	37	939.8	62	1574.8	87	2209.8
13	330.2	38	965.2	63	1600.2	88	2235.2
14	355.6	39	990.6	64	1625.6	89	2260.6
15	381.0	40	1016.0	65	1651.0	90	2286.0
16	406.4	41	1041.4	66	1676.4	91	2311.4
17	431.8	42	1066.8	67	1701.8	92	2336.8
18	457.2	43	1092.2	68	1727.2	93	2362.2
19	482.6	44	1117.6	69	1752.6	94	2387.6
20	508.0	45	1143.0	70	1778.0	95	2413.0
21	533.4	46	1168.4	71	1803.4	96	2438.4
22	558.8	47	1193.8	72	1828.8	97	2463.8
23	584.2	48	1219.2	73	1854.2	98	2489.2
24	609.6	49	1244.6	74	1879.6	99	2514.6
25	635.0	50	1270.0	75	1905.0	100	2540.0

The above table is exact on the basis: 1 in. = 25.4 mm

Conversion of millimeters to inches.

mm	in.	mm	in.	mm	in.	mm	in.
1	0.039370	26	1.023622	51	2.007874	76	2.992126
2	0.078740	27	1.062992	52	2.047244	77	3.031496
3	0.118110	28	1.102362	53	2.086614	78	3.070866
4	0.157480	29	1.141732	54	2.125984	79	3.110236
5	0.196850	30	1.181102	55	2.165354	80	3.149606
6	0.236220	31	1.220472	56	2.204724	81	3.188976
7	0.275591	32	1.259843	57	2.244094	82	3.228346
8	0.314961	33	1.299213	58	2.283465	83	3.267717
9	0.354331	34	1.338583	59	2.322835	84	3.307087
10	0.393701	35	1.377953	60	2.362205	85	3.346457
11	0.433071	36	1.417323	61	2.401575	86	3.385827
12	0.472441	37	1.456693	62	2.440945	87	3.425197
13	0.511811	38	1.496063	63	2.480315	88	3.464567
14	0.551181	39	1.535433	64	2.519685	89	3.503937
15	0.590551	40	1.574803	65	2.559055	90	3.543307
16	0.629921	41	1.614173	66	2.598425	91	3.582677
17	0.669291	42	1.653543	67	2.637795	92	3.622047
18	0.708661	43	1.692913	68	2.677165	93	3.661417
19	0.748031	44	1.732283	69	2.716535	94	3.700787
20	0.787402	45	1.771654	70	2.755906	95	3.740157
21	0.826772	46	1.811024	71	2.795276	96	3.779528
22	0.866142	47	1.850394	72	2.834646	97	3.818898
23	0.905512	48	1.889764	73	2.874016	98	3.858268
24	0.944882	49	1.929134	74	2.913386	99	3.897638
25	0.984252	50	1.968504	75	2.952756	100	3.937008

The above table is approximate on the basis: 1 in. = 25.4 mm, 1/25.4 = 0.039370078740+

Inch to metric equivalents.

Fraction (in.)	Decimal Equivalent		Fraction (in.)	Decimal Equivalent	
	English (in.)	Metric (mm)		Englilsh (in.)	Metric (mm)
1/64	.015625	0.3969	33/64	.515625	13.0969
1/32	.03125	0.7938	17/32	.53125	13.4938
3/64	.046875	1.1906	35/64	.546875	13.8906
1/16	.0625	1.5875	9/16	.5625	14.2875
5/64	.078125	1.9844	37/64	.578125	14.6844
3/32	.09375	2.3813	19/32	.59375	15.0813
7/64	.109375	2.7781	39/64	.609375	15.4781
1/8	.1250	3.1750	5/8	.6250	15.8750
9/64	.140625	3.5719	41/64	.640625	16.2719
5/32	.15625	3.9688	21/32	.65625	16.6688
11/64	.171875	4.3656	43/64	.671875	17.0656
3/16	.1875	4.7625	11/16	.6875	17.4625
13/64	.203125	5.1594	45/64	.703125	17.8594
7/32	.21875	5.5563	23/32	.71875	18.2563
15/64	.234375	5.9531	47/64	.734375	18.6531
1/4	.250	6.3500	3/4	.750	19.0500
17/64	.265625	6.7469	49/64	.765625	19.4469
9/32	.28125	7.1438	25/32	.78125	19.8438
19/64	.296875	7.5406	51/64	.796875	20.2406
5/16	.3125	7.9375	13/16	.8125	20.6375
21/64	.328125	8.3384	53/64	.828125	21.0344
11/32	.34375	8.7313	27/32	.84375	21.4313
23/64	.359375	9.1281	55/64	.859375	21.8281
3/8	.3750	9.5250	7/8	.8750	22.2250
25/64	.390625	9.9219	57/64	.890625	22.6219
13/32	.40625	10.3188	29/32	.90625	23.0188
27/64	.421875	10.7156	59/64	.921875	23.4156
7/16	.4375	11.1125	15/16	.9375	23.8125
29/64	.453125	11.5094	61/64	.953125	24.2094
15/32	.46875	11.9063	31/32	.96875	24.6063
31/64	.484375	12.3031	63/64	.984375	25.0031
1/2	.500	12.7000	1	1.000	25.4000

Metric equivalents.

Length

U.S. to Metric	Metric to U.S.
1 inch = 2.540 centimeters	1 millimeter = .039 inch
1 foot = .305 meter	1 centimeter = .394 inch
1 yard = .914 meter	1 meter = 3.281 feet or 1.094 yards
1 mile = 1.609 kilometers	1 kilometer = .621 mile

Area

1 inch^2 = 6.451 centimeter^2	1 millimeter^2 = .00155 inch^2
1 foot^2 = .093 meter^2	1 centimeter^2 = .155 inch^2
1 yard^2 = .836 meter^2	1 meter^2 = 10.764 foot^2 or 1.196 yard^2
1 acre^2 = 4,046.873 meter^2	1 kilometer^2 = .386 mile^2 or 247.04 acre^2

Volume

1 inch^3 = 16.387 centimeter^3	1 centimeter^3 = 0.61 inch^3
1 foot^3 = .028 meter^3	1 meter^3 = 35.314 foot^3 or 1.308 yard^3
1 yard^3 = .764 meter^3	1 liter = .2642 gallons
1 quart = .946 liter	1 liter = 1.057 quarts
1 gallon = .003785 meter^3	1 meter^3 = 264.02 gallons

Weight

1 ounce = 28.349 grams	1 gram = .035 ounce
1 pound = .454 kilogram	1 kilogram = 2.205 pounds
1 ton = .907 metric ton	1 metric ton = 1.102 tons

Velocity

1 foot/second = .305 meter/second	1 meter/second = 3.281 feet/second
1 mile/hour = .447 meter/second	1 kilometer/hour = .621 mile/second

Acceleration

1 inch/second^2 = .0254 meter/second^2	1 meter/second^2 = 3.278 feet/second^2
1 foot/second^2 = .305 meter/second^2	

Force

N (newton) = basic unit of force, kg-m/s^2. A mass of one kilogram (1 kg) exerts a gravitational force of 9.8 N (theoretically 9.80665 N) at mean sea level.

Temperature Conversion: Celsius (°C) to Fahrenheit (°F)

Temp. Celsius = 5/9 (Temp. °F − 32 deg.)

Temp. Fahrenheit = 9/5 × Temp. °C + 32 deg.

Rankine (Fahrenheit Absolute) = Temp. °F + 459.67 deg.

Kelvin (Celsius Absolute) = Temp. °C + 273.15 deg.

Freezing point of water: Celsius = 0 deg.; Fahr. = 32 deg.

Boiling point of water: Celsius = 100 deg.; Fahr. = 212 deg.

Absolute zero: Celsius = −273.15 deg.; Fahr. = −459.67 deg.

Appendix B

Tracing back over the years, it is found that 90% of fire fatalities or injuries on wildland fires can be directly attributed to a violation of the standard fire orders. Both the standard orders and the obvious or "watch out" situations can be printed in card form which can then be carried with the firefighter in a pocket or helmet.

Standard Fire Orders

- Keep informed of fire weather conditions and forecasts.
- Know what your fire is doing at all times.
- Base all actions in current and expect fire behavior.
- Plan escape routes for everyone and make them known.
- Post a lookout where there is possible danger.
- Be alert, keep clam, think clearly and act decisively.
- Maintain good communications at all times.
- Give clear instructions and be sure they are understood.
- Maintain control of your personnel at all times.
- Fight fire aggressively, but provide for safety first.

U.S. Forest Service, 2006

The following "watch out" situations were developed after a detailed study of many fire accidents, injuries, and deaths. These situations are a warning sign for firefighters to be extra careful as they are finding themselves in a situation where others have suffered.

Watch Out Situations

- You are given an assignment not clear to you.
- You cannot see the main body of the fire, and you are not in communication with anyone who can.
- You are getting spot fires over your line.
- You are attempting a direct attack of the fire.
- You are in an area where you do not know local fire behavior conditions.
- You are working in an area that you have not seen in daylight.
- You are working in steep, broken topography.
- You are working in direct attack in heavy fuels.
- Weather is getting hotter and drier.
- You notice a wind change.
- You notice rolling materials on the slope you are working on.
- You are assigned to construct a line downhill.
- You or your crew complains of headaches, fatigue or drowsiness while working.
- You are assigned to work on a line with dozer or air-attack resources.
- Your line assignment has chutes, chimneys, and a saddle to work across.
- You or your crewmembers are out of condition.

U.S. Forest Service, 2006

ACRONYMS

ADA   Americans with Disabilities Act

AFFF   Aqueous film-forming foam

ASTM   American Society for Testing and Materials

BLEVE   Boiling liquid expanding vapor explosion

BTU   British thermal unit

CAFS   Compressed air foam systems

CERCLA   Comprehensive Environmental Response Compensation and Liability Act

CHEMTREC   Chemical Transportation Emergency Center

CISM   Critical incident stress management

CMA   Chemical Manufacturers Association

CNG   Compressed natural gas

DHS   Department of Homeland Security

DOD   Department of Defense

DOE   Department of Energy

DOJ   Department of Justice

DOT   Department of Transportation

EEOC   Equal Employment Opportunity Commission

EMS   Emergency Medical Services

EPA   Environmental Protection Agency

EPCRA   Emergency Planning and Community Right-to-Know Act

ERG   *Emergency Response Guidebook*

ERT   Emergency response technology

FAA   Federal Aviation Administration

FBI   Federal Bureau of Investigation

FEMA   Federal Emergency Management Agency

FMZ   Fire management zone

FOG   Field operations guide

FRA   Federal Railroad Administration

HRR   Heat release rate

HSPD   Homeland Security presidential directive

HVAC   Heating, ventilation, and air conditioning

IAQ   Indoor air quality

IC   Incident commander

ICS   Incident Command System

IDLH   Immediately dangerous to life and health

IMO   International Maritime Organization

ISO   Insurance Services Office

JFO   Joint field office

JOC   Joint operations center

LEL   Lower explosive limit

LEPC   Local emergency planning committee

LNG   Liquefied natural gas

MBE   Management by exception

MBO   Management by objectives

MCI   Mass casualty incident

MDT   Mobile data terminal

MOU   Memorandum of Understanding

MSDS   Materials Safety Data Sheets

NBC   Nuclear biological chemical

NCP   National Oil and Hazardous Waste Pollution Contingency Plan

NDPO   National Domestic Preparedness Office

NFIRS   National Fire Incident Reporting System

NFPA   National Fire Protection Association

NHTSA   National Highway Traffic Safety Administration

NIMS   National Incident Management System

NIOSH   National Institute of Occupational Safety and Health

NIST   National Institute of Standards and Technology

NOAA   National Oceanic and Atmospheric Administration

NRC   Nuclear Regulatory Commission

NRT   National Response Team

NWCG   National Wildfire Coordination Group

OSHA   Occupational Safety and Health Administration

PAR   Personnel accountability report

PAS   Personnel accountability system

PASS   Personal alert safety system

PPE   Personal protective equipment

PPV   Positive pressure ventilation

RDD   Radiological dispersal device

RECEO-VS   Rescue, exposures, confinement, extinguishment, overhaul, and ventilation and salvage

REM   Roentgen equivalent man

RIC   Rapid intervention crew

RIT   Rapid intervention team

RIV   Rapid intervention vehicle

SARA   Superfund Amendment and Reauthorization Act

SCBA   Self-contained breathing apparatus

SERC   State emergency response commission

SIC   Side impact curtain

SIPS   Side impact protection system

SOLAS   Safety of Life at Sea (treaty)

SOP   Standard operating procedure

SRS   Supplemental restraint system

UC   Unified command

UEL   Upper explosive limit

USCG   U.S. Coast Guard

USFA   U.S. Fire Administration

VES   Vent, enter, search

WMD   Weapons of mass destruction

GLOSSARY

Absolute pressure The measurement of pressure exerted on a surface, including atmospheric pressure, measured in pounds per square inch absolute.

Absolute zero The temperature at which no movement of molecules occurs.

Access stairs The stairs that serve a particular floor or area of a building.

Acidity A substance that will release hydrogen ions when dissolved in water.

Aerial fuels Fuels that include all green and dead materials located in the upper forest canopy.

Air bills Papers identifying hazardous materials being shipped by air. The captain is responsible for maintaining control over the documents, which are stored in the cockpit.

Air stability and instability The relationship between the vertical temperature distribution within an air mass and a vertically moving parcel of air.

Alkalinity The amount of reaction of a substance with acids in aqueous solution to form salts, while releasing heat.

Alpha radiation The least dangerous external radiation hazard and the most dangerous internal radiation. External alpha particles have very low penetrating power and as a result seldom penetrate the skin. Turnout clothing will provide the needed protection from alpha radiation.

Alternative fuels Presently two types of compressed gases are being used for motor vehicle fuel. One is compressed liquefied natural gas (CNG) and the other is compressed liquefied petroleum (CLP) either propane or butane gas. Research is now being conducted to expand the use of hydrogen gas. These gases are under high pressure and present a danger when released.

America Burning The 1973 report from a committee appointed by the president to investigate and report back its findings regarding U.S. Fire Service. Over ninety recommendations were forwarded for action.

America Burning Recommissioned, America at Risk The 2000 revisit of both the original *America Burning* report (1973) and the *America Burning Revisited* report (1987). Two areas noted as needing further attention were fire prevention and firefighter safety.

America Burning Revisited The 1987 revisit to the *America Burning* report to evaluate the progress on accomplishing the needed improvements as recommended.

Anchor point An advantageous location, usually a barrier to fire spread, from which to start constructing fire line. This point is used to minimize the chance of being flanked by the fire while the line is being constructed, and may also refer to a safe ending point for a constructed line.

Aqueous film-forming foam (AFFF) A foam that is formed by a combination of water and perfluorocarboxylic acid. The surface tension of the fuel then exceeds the aggregate surface tensions of the film, preventing oxygen from mixing with the fuel vapors.

Area ignition The ignition of several individual fires throughout an area either simultaneously or in quick succession, and closely spaced so they influence and support each other to produce fast, hot spread of fire throughout the area.

Aspect (exposure) The direction in which the slope of a hill or mountain faces the sun.

Atom The smallest particle of a molecule, consisting of a neutron that is electrically neutral, a proton that has a positive electric charge, and a surrounding cloud of orbiting electrons that are negatively charged.

Auto-exposure The lapping of fire from one floor to the upper floor on the exterior of the building.

Auto-ignition temperature The minimum temperature to which a material must be raised before combustion will occur. Also called "ignition temperature."

Backdraft A sudden, violent reignition of the contents of a closed container fire that has consumed the oxygen within the space when a new source of oxygen is introduced. The introduction of oxygen results in an immediate smoke explosion.

Backfire The area between the control line and the fire's edge that is intentionally fired out (burned) to eliminate fuel in advance of the fire, and/or to change the direction of the fire and/or to slow the fire's progress.

Backpressure (nozzle reaction pressure) The pressure exerted in the opposite direction of the water flowing from a nozzle.

Balloon frame construction A type of construction in which the wood studs run from the foundation to the roof and the floors are nailed to the studs. This provides a wall space or channel for hot fire gases to spread vertically to the attic area. See *platform construction.*

Barometer An instrument for measuring atmospheric pressure.

Basic A substance that will react with acids in aqueous solution to form salts, while releasing heat.

Beta radiation Radiation that is given off by beta particles, which are smaller than alpha particles but have more penetrating power. Large amounts of beta radiation externally can seriously damage skin tissue. Beta particles enter the body through damaged skin by ingestion or inhalation and will cause organ damage similar to that caused by alpha particles. Full protective clothing including SCBA is required to protect against beta radiation.

Bilge The bottom area of a boat or ship where flammable liquids and vapors accumulate.

Bill of lading Documents indicating the amount and type of cargo or freight being transported.

Black zone The side of a fire line that is burned and used as the escape route for firefighters.

BLEVE The result of the explosive release of vessel pressure, portions of the metal tank, and a burning vapor cloud of gas with the accompanying radiant heat.

Blown-on (lightweight) concrete A lightweight concrete mixture that is blown onto metal to provide a coat of fire-resistive material. The fire-resistive rating varies with the type of material and its thickness.

Boiling point (BP) The temperature where a liquid will convert to a gas at a vapor pressure equal to or greater than atmospheric pressure.

Boil over The expulsion of a tank's contents by the expansion of water vapor that has been trapped under the oil and heated by the burning oil and metal sides of the tank.

Bond (chemical) The attractive force that often binds two atoms into a combination that is stable at least at room temperature, but which becomes unstable at a sufficiently high temperature.

Bottom (toe) The rear-most area of a fire, generally used as an anchor point or starting point for suppression activities.

Boyle's law States that the more a gas is compressed, the more the gas becomes difficult to compress further.

Boxcar A railroad car that is enclosed and used to carry general freight.

British thermal unit (BTU) The amount of heat required to raise the temperature of one pound of water one degree Fahrenheit.

Brush Shrubs and stands of short scrubby tree species that do not reach merchantable size.

Bulkhead A main wall or supporting structure of a ship, generally constructed of steel. Some bulkheads are watertight and fire resistive. They can be designed and constructed to prevent the spread of water and fire, if the ship has been damaged.

Burning index A number in an arithmetic scale determined from fuel moisture content, wind speed, and other selected factors that affect burning conditions and from which the ease of ignition of fires and their probable behavior may be estimated.

Carbon monoxide (CO) The result of the incomplete combustion of a fuel where carbon and a single atom of oxygen is produced.

Catalyst A substance that greatly affects the rate of a chemical reaction, but is not created or destroyed in the chemical reaction. The catalyst helps to lower the energy activation thus increasing the reaction rate.

Charles's law States that a gas will expand or contract in direct proportion to an increase or decrease in temperature.

Chemical change The molecules of the materials are changed with the process.

Chemical Manufacturers Association (CMA) An association of the manufacturers of hazardous chemicals, which funds CHEMTREC.

CHEMTREC (Chemical Transportation Emergency Center) A national communications center that provides technical advice and guidance for emergency responders at the scene of a chemical emergency. This network of chemical manufacturers provides emergency information and response teams if necessary.

Chimneys Steep, narrow draws in canyons. Radiant or convective heating or spotting can occur across narrow canyons due to the short distances involved.

Churning of the air The phenomenon of smoke being blown out at the top of the opening, only to be drawn back into the structure at the bottom of the opening by the slight negative pressure (vacuum) created by the action of the mechanical fan.

Cell(s) The composition of all organisms, where all vital functions of the organism occur within its walls.

Cellulose materials A complex carbohydrate of the cell walls of plants used in making paper or rayon.

Central (center) core construction A type of building construction in which elevators, stairs, and support systems are located in the center of the building.

Class A fires Fire involving ordinary combustibles.

Class B fires Fires involving flammable liquids.

Class C fires Fires involving energized electrical equipment or wires.

Class D fires Fires involving combustible metals.

Class K fires Fires involving cooking oils.

Class A foam Foam intended for use on Class A fires.

Class B foam Foam intended for use on Class B fires.

Closed cup test Place a lid on the cup to confine the vapors above the cup. The point vapors are given off in sufficient quantities to ignite this point is called a *closed cup reading*. The ignition point is always lower in a closed cup test than an open cup test. The devices used to perform this test are the Tag Closed Cup Test for liquids below 200°F (93°C) and the Pensky-Martens Closed Cup Apparatus for liquids above 200°F (93°C).

Cockloft The void space, approximately 3 ft deep, between the ceiling area and the underside of the roof. This area is important to firefighters because the space allows heated fire gases to move and spread fire into other parts of the building.

Cold trailing A method of making sure the fire is out by carefully inspecting and feeling with the hand the dead edge of the fire to detect any live fire.

Collapse zone The area 1.5 times the height of the building in which no personnel are allowed to operate as the building may collapse. Unmanned master streams are generally placed within this zone.

Combination attack A method that takes advantage of the direct attack and the indirect attack when possible by combining both methods simultaneously. A direct attack is used on areas of the fire line that can be confronted safely, while areas that are considered too dangerous for this method are attacked using indirect methods.

Combustion A chemical process between fuel and oxygen with the evolution of light either as a glow or flame and heat. Some of the heat energy is radiated back into the fuel, releasing more fuel to allow the combustion process to continue.

Compartmentation A fire occurring in an enclosed area that restricts the movement of fire gases and the consumption of oxygen and traps the by-products of the combustion process.

Compound A substance formed from two or more elements joined with a fixed ratio.

Compressed air foam system (CAFS) A technology developed for foam-producing systems generally installed on the fire engine or special unit. The amount of wetness or dryness of the foam can be set. Once programmed, the special unit automatically regulates the amount of water, foam concentrate, and air pressure. It manufactures a uniform mixture of foam bubbles.

Compressed gas Any material that, when enclosed in a container, has an absolute pressure of more than 40 psi at 70°F, or an absolute pressure exceeding 104 psi at 130°F, or both.

Compressed natural gas (CNG) Natural gas that is compressed and contained within a pressurized container. It provides the fuel to power some of the new hybrid vehicles and many public buses.

Conduction The transfer of heat energy by the movement of the heat-agitated atoms colliding with each other, transmitting some of the energy or heat.

Confinement An activity required to prevent fire from extending to an uninvolved area or another structure.

Conflagration A fire with major building-to-building flame spread over a great distance.

Consist (way) bill Documents used to identify the hazardous cargo of a rail transportation vehicle. The driver of the vehicle has the responsibility to maintain these documents, which are stored in the cab of the vehicle.

Control valves Valves used to control the flow of water in sprinkler or standpipe systems.

Convection The movement of heat energy by the agitation of air molecules reduces the density of molecules making heated air lighter than cooled air. In a heated enclosed compartment, the heated air rises and pulls in cooled air below the flaming level.

Convection column The ascending column of gases, smoke, and debris produced by the combustion process or a fire.

Cribbing Blocks of wood or prefabricated plastic blocks used to lift a vehicle's frame off the wheels and stabilize the vehicle.

Crown fires A fire that advances across treetops and shrubs more or less independently of the surface fire. Sometimes crown fires are classed as either running or dependent, to distinguish the degree of independence for the surface fire.

Cryogenic Gaseous materials with boiling points of no greater than −150°F that are transported, stored, and used as liquids.

Cybernetic building systems Automatic control systems installed in buildings to integrate the building services in a central location within the building, including energy management (HVAC systems), fire detection and security systems, building transportation systems, fault detection, and diagnostics.

Dangerous cargo manifest Documents used to identify the hazardous cargo of a marine vessel. The captain or shipmaster has the responsibility to maintain these documents.

Decay stage (burnout phase) The stage at which fire has consumed all the available fuel, and the temperature begins to decrease as the fire reduces in intensity.

Decision-making model A five-step process used to solve problems: (1) Identify the problem(s); (2) identify the alternatives available; (3) select the best alternative; (4) implement the chosen solution; and (5) evaluate the implemented solution for expected results.

Defensive mode Generally a fire strategy conducted on the exterior of the building to protect buildings adjacent to the building on fire from spreading into the unaffected buildings.

Deluge system A system designed to protect areas that may have fast-spreading fire engulfing the entire area. All of its sprinkler heads are open, and the piping contains atmospheric air. Upon system operation, water flows to all heads providing total water coverage. The system has a deluge valve that opens when activated by a separate fire detection system.

Department of Defense (DOD) One of the federal agencies responsible for supplying manpower, training, and equipment worldwide for a full range of hazardous substances. The DOD laboratories and bases can be a source of expertise, equipment, and supplies for use in local emergencies.

Department of Energy (DOE) The federal agency responsible for radioactive waste generated by the nuclear weapons program or by the nuclear reactors supplying energy.

Department of Homeland Security (DHS) The federal agency responsible for protecting U.S. territory from terrorist attacks and for responding to natural disasters.

Department of Labor (DOL) The federal agency responsible for promoting and protecting wage earners with the use of its agency, the Occupational Health and Safety Administration, to provide regulations and enforcement to ensure that occupational environments are safe for workers.

Department of Transportation (DOT) The federal agency responsible for issuing standards and regulations relating to the transportation of hazardous materials nationwide. The DOT also trains and inspects carriers and shippers of hazardous materials to ensure that they are in full compliance.

Diffusive flaming A combustion process where the flames are part of the actual combustion process.

Direct method of attack Attack that is directed to the edge of the burning fire.

Drafting The process of moving or drawing water away from a static source of water by a pump.

Drain time The time it takes for the foam to break down as the water separates (drains) from the aerated foam bubbles when they break.

Drywall See *gypsum board*.

Electron A very light particle with a negative electric charge, a number of which surround the nucleus of most atoms.

Element The simplest form of matter.

Emergency Planning and Community Right-to-Know Act (EPCRA) An act that encourages and supports emergency planning efforts at the state and local levels and provides the public and local governments with information concerning potential chemical hazards present in their communities.

Emergency Response Guidebook (ERG) This booklet contains the chemical names and the commonly used names for most hazardous chemicals. It also has an identification numbering system as well as a placard system for transportation vehicles and containers.

Endothermic The type of reaction in which energy is absorbed when the reaction takes place.

Entrain (entrainment) The gathered or captured cooler air as the heated air surrounding the point of combustion rises.

Envelopment action Taking a suppression action on a fire at many points and in many directions simultaneously. It provides for a rapid attack and, if coordinated properly, can be highly effective on smaller fires.

Environmental Protection Agency (EPA) A federal agency concerned with materials that are hazardous to the environment.

Evaporation The process of air moving over the surface of water while picking up water vapor. Water vapor is simply water in a gaseous state, and is an important ingredient in fire behavior.

Exothermic A type of reaction that releases or gives off energy.

Explosive range The range of concentrations of the gases or materials (dusts) in the air, which will permit the material to burn.

Exposure Property that may be endangered by radiant heat from a fire in another structure or an outside fire. Generally, property within 40 ft is considered an exposure risk, but larger fires can endanger property much farther away.

Exposure See *aspect*.

Eyebrow(s) A concrete extension over the openings on a multistory building to prevent the expansion of fire from lower rooms to the upper floors.

Federal Aviation Administration (FAA) The federal agency responsible for regulating and overseeing all aspects of U.S. civil aviation.

Federal Bureau of Investigation (FBI) The federal agency responsible for crime investigation at acts of terrorism, under direction of Homeland Security.

Federal Emergency Management Agency (FEMA) The federal agency responsible for consequence management.

Field operations guide (FOG) A small notebook designed to be carried on person, which contains complete data on the ICS, including job titles, descriptions, and equipment classifications systems.

Fingers of a fire The long, narrow tongues of a fire projecting from the main body.

Fire A rapid, self-sustaining oxidation process that involves heat, light, and smoke in varying quantities. It is often an unplanned or uncontrolled event, as in most cases the fuels are not selected or even known in advance. See also *combustion*.

Fire dampers Equipment used where ducts penetrate rated walls, floors, or partitions. They are installed inside at the point of penetration to preserve the integrity of the fire area. They also can be used inside ducts to prevent backdrafts, keeping undesirable vapors and dusts from being moved upstream from their collection point in the systems.

Fire load   The total amount of fuel that might be involved in a fire, as measured by the amount of heat that would evolve from its combustion, expressed in BTUs. See also *BTU*.

Fire management zone (FMZ)   A zone within a jurisdictional engine company's area where similar hazards are grouped by approximately equal needed fire flow and hazard. Once they are grouped, a hazard severity ranking is given to them. Manpower and equipment response can be established using the hazard severity ranking given to each FMZ.

Fire point   The lowest temperature at which a liquid produces a vapor that can sustain a continuous flame, opposed to the instantaneous flash as described in the flash point. See also *flash point*.

Fire pump   A specially designed and listed pump that increases the pressure of the water serving a fire protection system.

Fire retardant   Any substance that by chemical or physical action reduces flammability of combustibles.

Fire tetrahedron   A model now replacing the "old fire triangle" as a better description of the combustion process. The combustion process needs four things to continue burning: heat, fuel, oxygen, and chemical reaction.

Fire triangle   A three-sided figure depicting heat, fuel, and oxygen. The model demonstrates that when one side of the triangle is removed, the combustion process will stop.

Firing out   Used during a direct fire attack to burn out fuel between the main fire line and the line being constructed for fire control.

Flame over (rollover)   Flames traveling through or across the unburned gases in the upper portions of the confined area during the fire's development.

Flaming combustion   An exothermic or heat-producing chemical reaction with flames occurring between some substance and oxygen.

Flammable gas   A gas that is flammable at atmospheric temperature and pressure in a mixture of 13% or less (by volume) with air and has a flammable range wider than 12% regardless of the lower limit.

Flammable range   The numerical difference between a flammable substance's lower and upper explosive limits in air.

Flanking action   The fire is attacked on both flanks at the same time, with the objective to prevent the fire from spreading on a given flank and thus threatening some exposure such as heavier fuel, a recreational area, or an area of structures.

Flanks of a fire   The parts of a fire's perimeter that are roughly parallel to the main direction of spread.

Flash fuels   Grass, leaves, ferns, tree moss, and some kinds of slash that ignite readily and are consumed rapidly when dry.

Flashover fire   A sudden event that occurs when all the contents of a room or enclosed compartment reach their ignition temperature almost simultaneously resulting in an explosive fire.

Flash point   The temperature of a liquid which, if an ignition source is present, will ignite only the vapors being produced by the liquid creating a flash fire. See also *fire point.*

Flatcar   A railcar that is open to the elements.

Foehn   (pronounced "fane") A dry wind with a strong downward component locally called Santa Ana, North, Mono, Chinook, or East Wind.

Free radicals   A molecular fragment possessing at least one unpaired electron. It is very active chemically and must react with another free radical to form a compound.

Friction loss   The pressure lost by fluids while moving through pipes, hose lines, or other limited spaces.

Fuel-controlled fire   A fire controlled by the amount and type of fuel consumed.

Fuel loading   A term that refers to the amount of fuel available to burn in a given area.

Fuel type   An identifiable association of fuel elements of distinctive species, form, size, arrangement, or other characteristics that will cause a predicable rate of fire spread or difficulty of control under specified weather conditions.

Fully developed stage   Stage or phase of a fire where the maximum generation of heat and consumption of fuel and oxygen occurs.

Functional groups   All of the elements of a group having the same electronic structure in the outermost shell, and thus showing similar chemical behavior.

Gamma radiation   The most dangerous form of penetrating radiation. Gamma rays have 10,000 times the penetrating power of alpha particles and 100 times the penetrating power of beta particles. They pass through the body and strike body organs and in doing so cause extensive internal damage.

Global positioning   Using satellites positioned in outer space locations on earth can be triangulated by the beam from several satellites; this triangulation allows accurate identification of a position on the earth to be determined. Fire mapping systems can be integrated into the GPS for computer modeling on wildland fires and for tracking engine companies available for response.

Gondola cars   Train cars designed with flat bottoms and four walls and can carry a variety of goods. They may have a cover for the cargo being carried. A few are constructed of wood, but most often are constructed of steel.

Gravity winds   Winds that occur when air is pushed over high elevations and flows downhill. They are identified in a number of ways including foehn winds, devil winds, sundowners, Chinook winds, and a variety of other names depending on their location.

Green area   The area next to a vegetation fire that has not been burned. Firefighters always watch for and protect against controlled fires moving into the green.

Growth stage   Where the fire increases its fuel consumption and heat generation.

Gypsum board   Used in interior finishes for walls and ceilings, made of a core of calcinated gypsum, starch, water, and other additives that are sandwiched between two heavy paper sheets. Also called "drywall" or "plasterboard."

Halogenation (halogenated)   Any chemical reaction in which one or more halogen atoms are incorporated into a compound.

Hazardous materials   Any substance or material in any form or quantity that poses an unreasonable risk to safety, health, and property when transported in commerce (DOT 49 CFR 171).

Hazardous substances   Any substance designated under the Clean Water Act and the Comprehensive Environmental Response Compensation and Liability Act (CERCLA) as posing a threat to waterways and the environment when released. When used by OSHA, hazardous substances refers to any substance defined by EPA, within section 101 of CERCLA; or as any biological agent or other disease-causing agent as defined by EPA within section 101 of CERCLA; or as any substance listed by DOT as a hazardous material and any hazardous waste as defined by EPA in 40 CFR 261.3 or by DOT in 49 DFR 171.8.

Head   The outermost portion of the fire that is moving from the rear.

Heat   All of the energy, both kinetic and potential, within molecules.

Heating, ventilation, and air conditioning (HVAC) system   A central system used to heat and cool large buildings. They are of concern to firefighters as they can spread hot fire gases, if not equipped with fire dampers to control the movement of the heated gases and smoke.

Heat release rate (HRR)   The rate at which heat is generated by a source, usually measured in watts, joules/sec, or BTUs/sec.

Heavy timber construction   A type of building construction in which the exterior walls are usually made of masonry, and therefore, are noncombustible. The interior structure members are constructed with large, unprotected, wood members. To meet this definition, the columns must be at least 8 inches square and beams must be at least 6 by 10 inches.

Hopper car   A type of railroad freight car used to transport loose bulk commodities.

Hybrid passenger vehicles   A term used to describe vehicles powered by a variety of fuels and powering systems. Hydrogen, compressed natural and liquefied petroleum gases, and preheated vegetable oil can be found. The powering systems in some use higher voltage systems up to 600 volts. Some vehicles combine gas turbines and a battery system. Most are experimental and a number have resulted in fires. New hydrogen fueling stations are being designed and built. Firefighters need to stay on top of the research and development of products in this area.

Hydrocarbon products Organic compounds, such as benzene and methane, that contain only carbon and hydrogen.

Incident Command System (ICS) A standardized on-site emergency management concept specifically designed to allow user(s) to adopt an integrated organizational structure equal to the complexity and demands of single or multiple incidents, without being hindered by jurisdictional boundaries. The command portion is only one component of the newer, more comprehensive National Incident Management System (NIMS).

Incipient (ignition) stage The point at which the four components of the fire tetrahedron come together and the materials reach their ignition temperature and the fire begins.

Indirect attack method Consists of fighting a wildland fire by constructing control lines or by backfiring the fuel out at a distance ahead of the main fire.

Inflator A pressurized container which holds gas that is released upon impact, filling the curtain or bag which restrains passengers within the seating area. The container in the newer hybrid cars can contain up to 3,000 psi.

Infrared imaging A method using infrared waves to detect heat being radiated by a substance or body. Firefighters use a camera to translate the location and intensity of the heat being given off. Using these cameras they can locate victims or other firefighters who may be trapped.

Inhibitor Chemicals that are added in small quantities to unstable materials to prevent a vigorous reaction. Also called a "stabilizer."

Inorganic A classification that indicates matter is comprised chiefly of the materials of the earth, such as rocks, soil, air, water, and minerals in or below the surface of the earth.

Insolubility This term indicates the amount of a material that will dissolve and mix in water. Insoluble or slightly soluble materials will form a separate layer and will either float or sink, depending on their specific gravity.

Instability A condition of the atmosphere on which the lapse rate is such that a parcel of air given an initial vertical impulse will tend to move from its original level with increasing speed.

Insurance Services Office (ISO) An agency funded by the insurance companies to independently apply the grading schedule to cities (fire departments) and set the rate for fire insurance premiums.

Interface lands The areas where structures interface with wildland vegetation.

Intermodal railcar Transportation of freight in a container or vehicle, using multiple modes of transportation (rail, ship, and truck), without any handling of the freight itself when changing modes.

International Maritime Organization (IMO) An agency of the United Nations dealing with maritime issues. It is responsible for maintaining the SOLAS treaty.

Inversion layer A layer of comparatively warm air overlaying cool air. This is stable atmospheric condition. The atmosphere in an inversion will resist vertical motion of the air.

Islands Areas inside a wildland fire that have not been burned out.

Latent heat of vaporization The quantity of heat absorbed by a substance when it changes from a liquid to a vapor.

Law of latent heat of vaporization States that heat is absorbed when one gram of liquid is transformed into vapor at the boiling point under one atmosphere of pressure and the result is expressed in BTUs per pound or calories per gram.

Leeward side The side of the building the wind is blowing toward. See also *wayward side.*

Liquefied petroleum gas (LPG) A term given butane and propane gas that has been pressurized and contained in a tank. Both gases are heavier than air and have higher heat content than natural gas. They are widely used in recreational vehicles and can be found in homes in the more rural areas where natural gas or (methane) is not available.

Live load Items inside a structure that are not attached or of a permanent nature.

Local emergency planning committee (LEPC) Responsible for developing and maintaining a local emergency response plan to ensure quick and effective response to chemical emergencies. The plan provides effective actions to be taken and methods of evacuation if needed.

Louvering A method of roof ventilation that reduces the exposure to smoke and initially prevents cutting the supporting roof joists. It is especially effective on plywood-covered roofs.

Lower explosive limit (LEL) The lowest ignitable concentration of a substance in air that will ignite.

Manifest The listing of onboard cargo, possibly indicating the storage location of the various materials being shipped.

Mass A measurement of quantity when the weight is proportional to the mass.

Materials Safety Data Sheets (MSDS) A required worksheet prepared by manufacturers and marketers to provide workers with detailed information about the properties of a commercial product that contains hazardous substances.

Matter Anything that occupies space and has mass or something that occupies space and can be perceived by one or more senses.

Meter head (drip loop) The location where the electrical supply lines enter the building. A circuit breaker or electrical shutoff is generally found below the meter head at the circuit breaker box.

Molecule Two or more atoms tightly bound together by chemical bonds.

Monomers Small molecules that are usually gaseous or liquid and are used to produce polymer resins.

Nacelle The covering or cowling that protects the engine of the aircraft.

National Domestic Preparedness Office (NDPO) The coordination center for all federal efforts in response to weapons of mass destruction (WMD).

National Fire Protection Association (NFPA) 704 system The procedure of assigning one of the five numbers, 0 through 4, to the relative degree of three hazards, which are color coded (health, flammability, and chemical reactivity) for a given hazardous substance. A diamond-shaped figure is divided into four quadrants, each of which is color coded for each of three hazards (blue for health hazards, red for flammability, and yellow for chemical reactivity) and numbered 0 to 4 to designate the degree of the relevant hazard. The number 0 indicates that the materials are considered to be stable even in the event of fire.

National Highway Traffic Safety Administration (NHTSA) A group that developed a small-scale regulatory fire test for vehicle interior materials in the late 1960s and has continued applying the same test today.

National Incident Management System (NIMS) The newer comprehensive system designed to manage large incidents of significance. The system includes command and management (ICS functions), preparedness, resource management, communications and information management, supporting technologies, and ongoing management and maintenance.

National Institute of Occupational Safety and Health (NIOSH) An institute that provides toxicity testing and standards, publishes the *Guide to Chemical Hazards,* an important source of information for first responders, and is working to develop protective breathing and clothing standards for the weapons of mass destruction.

National Oceanic and Atmospheric Administration (NOAA) An organization that works with the EPA to develop a chemical reactivity worksheet that may be useful to first responders.

National Response Team (NRT) A team of fourteen federal agencies assembled to assist state and local agencies with hazardous materials incidents and the preplanning for such events.

Negative ion Forces outside an atom cause it to gain an extra electron, giving it an overall negative charge. It now contains more negative electrons than protons, so the sum of the electrical charge is negative.

Negative pressure ventilation A method of forced ventilation that pulls air and smoke out of a structure.

Neutron A particle with nearly the same mass as the proton, but electrically neutral; it is part of the nucleus of all atoms except the most common isotope of hydrogen.

Nonattack mode Under certain circumstances a fire attack may be too dangerous and the incident command will choose to "let the fire burn out" without a direct attack on it. However, exposures may be protected at a safe distance. See also *passive approach*.

Nuclear Regulatory Commission (NRC) A commission within the Department of Energy that recommends and establishes controls by providing licenses to facilities that use nuclear products.

Occupancy or use The building code that classifies buildings by the use of the building. Using this classification, the type of construction, size, and other fire and safety requirements are set forth in the appropriate codes.

Offensive mode Fire fighting operations that make a direct attack on a fire for purposes of control and extinguishment. For structural fire fighting situations, this usually means interior fire fighting.

Open cup test Flammable liquids are heated until the surface tension of the liquid is broken. At this breaking point, the fuel releases molecules in the form of vapor above the surface. The open cup test results in a higher ignition point than the closed cup test. The device used to perform this open cup test is called the Cleveland Open Cup Apparatus.

Ordinary construction A type of building construction in which the exterior walls are usually made of masonry and therefore noncombustible. The interior structural members may be combustible or noncombustible.

Organic A classification that generally means matter is found in living substances as plants and animal life.

Overhaul A systematic process of searching the fire scene for possible hidden fires or sparks that may rekindle. Overhaul is also used to assist on determining the origin and cause of the fire.

Oxidizing agents Substances that present special hazards to firefighters because they react chemically with large numbers of combustible organic materials such as oils, greases, solvents, paper, cloth, and wood. The halogens are also powerful oxidizers.

Parallel method A method of suppression in which fire line is constructed approximately parallel to and just far enough from the fire edge to enable personnel and equipment to work effectively, though the line may be shortened by cutting across unburned fingers. The intervening strip of unburned fuel is normally burned out as the control line proceeds, but may be allowed to burn out unassisted if there is undue threat to the line.

Particulates The unburned products of combustion, visible in smoke.

Parts of a (wildland) fire In typical fire burning fires the spread is uneven, with the main spread moving with the wind or upslope. The most rapidly moving portion is designated the head of the fire, the adjoining portions of the perimeter at right angles to the head are known as the flanks, and the slowest moving portions are known as the rear.

Passive approach On some wildland fires, structures may be bypassed by firefighters as the structure is fully involved and fire fighting efforts would be wasted on this particular structure. Another structure nearby may be only threatened and the resources could be used to save the threatened structure. Also see *nonattack mode*.

Personal alert safety system (PASS) A small, motion-sensitive unit attached to and worn with the SCBA by firefighters when entering an IDLH environment. As long as the firefighter moves, the alarm does not sound. If movement stops for thirty seconds, the device sends a "chirping" warning sound; if the wearer fails to move, the device will go into alarm mode and emit a loud noise to signal that the wearer may be in trouble.

Personnel accountability report (PAR) Some departments are required to report, generally by radio, when the situation deteriorates or some departments routinely require periodic reports on the location and condition of personnel who have entered the working area of an emergency incident.

Personnel accountability system (PAS) A system set up to determine the entry and exit of personnel into the working area of an emergency incident. A number of methods are used by the fire service.

pH   A measure of a substance's ability to react as an acid (low pH) or as an alkali (high pH).

Physical change   Molecules of the materials are not changed by the process.

Pike pole (rubbish hook)   A device with a sharp pointed head and a curved hook which is used to thrust into the ceiling area and then using the hook pull the ceiling materials downward. Because the hook is attached to a long pole, firefighters use it to ensure that all materials are removed when venting a roof from topside. The pike pole can also be used to sound the roof.

Piloted ignition temperature   The ignition temperature of a liquid fuel. When heated, it will self-ignite.

Pincer action   Moving crews along both flanks of the fire to a point where the flanking forces move closer together in a pinching action near the head of the fire.

Platform construction   A type of construction in which the floors in a building are built separately, thus in the outer walls the top of the ceiling and floor area serves as a fire block to stop the movement of hot fire gases between floors. See *balloon frame construction*.

Plenum space   A space above the suspended ceiling or hallway that is kept under negative pressure for return air.

Polar solvent   A substance that allows fire fighting foam to be used on alcohol-based fires without breaking down the soap-based materials in the foaming agent.

Polymerize   The process in which molecules of a monomer are made to combine with other monomers. Sometimes the reaction is explosive in nature.

Positive ion   An atom that is missing electrons.

Positive pressure ventilation   A process that uses mechanical fans to blow air into a structure to remove smoke and gases.

Postfire conference/postincident size-up   Following the incident, this analysis is conducted to exchange what was observed, the result of actions taken, and what lessons were learned so improvements can be made on future incidents.

Preemergency inspection   The fact-finding part of the preemergency planning process in which the facility is visited to gather a comprehensive list of information regarding the building and its contents.

Preincident planning   A process of preparing for operations at the scene of a given hazard or occupancy.

Pressure relief valve   Used on compressed gas cylinders to release pressure buildup within a cylinder. Generally, the pressure at which they open is preset.

Problem-solving process   See *decision-making model*.

Proton   A positively charged particle that is the nucleus of the most common isotope of hydrogen.

Pyrolysis   The process of breaking down a solid fuel into gaseous components when heated; also called thermal decomposition.

Rad   A measure of the dose of ionizing radiation absorbed by body tissues stated in terms of the amount of energy absorbed per unit of mass of tissue.

Radiation   The combined process of emission, transmission, and absorption of heat energy traveling through electromagnetic waves from an area of higher heat energy to an area of lower heat energy. A good example is the transfer of solar heat from the sun to the earth. The heat energy is moved through waves from an area of higher temperature (sun) to an area of lower temperature (earth).

Radiological dispersal device (RDD)   A device such as an explosive used to disperse radiological products such as fission products which are the waste from fuel of nuclear reactors or medical, industrial, and research radiological waste.

Rapid intervention team or crew (RIT/RIC)   The assignment of a group of resources with the sole purpose of rapid deployment to reports of operating personnel in trouble or missing. See NFPA 1710 and 1720, and FEMA technical report 123.

Rapid intervention vehicle (RIV)   The aircraft RIV is one example designed to rapidly deploy firefighters. For example, Los Angeles County has smaller and faster pumpers that carry limited water. The unit is able to respond rapidly in an effort to extinguish fires before they grow in size.

Rear (of the fire)   The portion of the edge of a fire opposite the head.

RECEO-VS An acronym used when developing the strategy and tactics on the fire ground. It stands for *rescue, exposures, confinement, extinguishment, overhaul, ventilation,* and *salvage.* The first five are listed in order of priority; the last two may be used at any point to support the first five.

Red flag alert The condition that occurs when relative humidity is below 15%, causing fuel to become very dry and thus making conditions favorable for a wildland fire. Fire departments then prepare for wild fires by moving and adding additional resources and by activating special fire prevention activities.

Reducer A substance used to reduce another substance by donation of electrons in the outer ring. Reducing agents need to be protected from air because they react with oxygen.

Rehabilitation A group of activities that ensures the health and safety of responders at emergency incidents. The activities include rest, medical care, hydration, and nourishment on an as-needed basis.

Rekindle Continuation of a fire generally after the fire department has declared the fire extinguished and left the scene.

Relative biological effectiveness (RBE) A measure of how damaging a given type of particle is when compared to an equivalent dose of x-rays.

Relative humidity The ratio of the amount of moisture in a given volume of space to the amount that volume would contain if it were saturated.

Rem A measure of the dose of ionizing radiation to body tissue stated in terms of its estimated biological effect relative to a dose of one roentgen (r) of x-rays.

Ridges Elements that divide the terrain. Different wind conditions may be present on each side of the ridge.

Risk assessment Activities that involve the evaluation or comparison of risks and the development of approaches that can change the probability or consequences of an incident.

RIT (RIC) See *rapid intervention team.*

Saddles The low topography between two high points.

Safety of Life at Sea (SOLAS) An international treaty containing minimum standards of fire and related

safety issues for ships on international voyages. The United States is a signatory to the SOLAS treaty; therefore, U.S. registered (flag) ships using international waters must comply with its provisions.

Salvage Described as the protection of property which was not damaged by the fire or could be damaged by the fire, smoke, water, and related extinguishment activities.

Saponification The process of chemically converting the fatty acid contained in the cooking medium to soap or foam.

Scuttle hole An opening that allows entry into the attic area. The cover during ventilation in some cases can be removed to ventilate the attic below.

Self-contained breathing apparatus (SCBA) Air is carried in a pressurized tank and provided to the person under a positive pressure within an enclosed mask.

Self-oxidizing materials Materials containing extra oxygen which enhances the process of combustion by intensifying the fire.

Siamese connection A hose fitting designed with a double hose line connection that allows two hose lines to be connected into one (generally larger) line.

Side impact curtain (SIC) A new vehicle passenger restraint system that deploys more quickly and forcefully from the side because of the shorter distance to the passenger.

Size-up A method used by firefighters to identify the problem(s) presented by the incident. Once the problems are identified, size-up procedures are used to determine the best overall strategy to be used to solve the problem. The tactics are those actions needed to complete the strategy or plan.

Slippery water Using polymers, a plasticlike additive to the water, it was found that the polymers not only reduced the friction loss in the hose but also increased the amount of water that could be moved through a hose line.

Smoke ejectors Mechanical smoke fan designed to draw heated air and smoke outside of a structure.

Smoldering (glowing) combustion The absence of flames with the presence of hot materials in the surface where oxygen diffuses into the surface of the fuel.

Soffits False spaces in the underside of a stairway and projecting roof eaves or the false space above built-in cabinets generally in kitchens or bathrooms.

Solubility A measure that indicates the tendency of a chemical to dissolve evenly in a liquid.

Sounding a roof Using an axe or pike pole, firefighters search the roof ahead of them to determine if it is solid enough to handle the additional weight of firefighters.

Specific gravity The density of the product divided by the density of water, with water defined as 1.0 at a certain temperature.

Spontaneous combustion An occurrence where a material self-heats to its piloted ignition temperature, then ignites.

Spot fires Fires that jump over a control line and start the vegetation burning ahead of the main fire front.

Stack effect The temperature difference between the outside temperature of the building and the temperature inside the building. This temperature difference creates air movement inside the building, which may carry heated fire gases to the upper or lower floors.

Stairwell support procedure A procedure used to move resources to the fire attack area when the elevators are unsafe to use. Firefighter teams position themselves two floors apart in the stairwell to carry resources two floors and return to rest and carry additional resources again. This relay system provides a steady flow of resources to the fire area.

Standard operating procedure (SOP) Specific information and instructions on how a task or assignment is to be accomplished.

Standpipes A manual fire fighting system with piping and hose connections inside buildings. They are classified into three categories: *Class I system* used for manual fire fighting with a 2.5-inch (64-mm) hose connection, *Class II system* used for manual fire fighting with a 1.5-inch (38-mm) hose connection but as a first-aid application only, and *Class III system* used for manual fire fighting with a combination of 2.5-inch (63.5-mm) and 1.5-inch (38-mm) connections.

State emergency response committee (SERC) Appointed by the governor, this committee is responsible for the coordination of emergency plans, which are developed by the LEPCs.

Stoichiometric An ideal burning situation or a condition where there is perfect balance between the fuel, oxygen, and end products.

Stoichiometry of reaction The proportion of fuel to oxygen and the resulting end products.

Stratification location The location in a high-rise building where light heated air flows upward and reaches a point where it is the same temperature and weight as the surrounding air, and at this location it flows horizontally.

Sublimation The changing from a solid to a gas directly without changing into a liquid.

Supplemental restraint system (SRS) An additional passenger restraint system, some of which are placed in the vehicle roof, dash, or in a roof pillar.

Supply diffuser The component of the HVAC system that distributes fresh air to a room; the return diffuser component draws stale air from the room.

Surfactant A soap material that works to ease the surface tension of water, allowing the water to more easily penetrate materials and thus increasing the effectiveness of water as an extinguishing agent.

Swamper A worker on a bulldozer crew who pulls the winch line, helps maintain equipment, and generally speeds suppression work on a fire.

Synergistic The aspect of two substances that interact to produce an effect that is greater than the sum of the substance's individual effects of the same type.

Tank railcar A tank mounted on a railroad frame with wheels designed to transport a variety of liquid products, having tank capacities ranging from a few hundred gallons to as much as 45,000 gallons (over 170 kiloliters).

Temperature A measure of the average molecular velocity or the degree of intensity of the heat.

Temperature inversion An action that occurs when cooler air is found close to the ground surface and warmer air is above this pool of cooler air.

Thermal imbalance A condition that occurs through turbulent circulation of steam and smoke in the fire area and leads to decreased visibility and uncomfortable conditions.

Thermal imaging device A device that distinguishes between objects or areas with different temperatures and displays hot areas and cold areas as different colors on a video screen.

Thermal stratification/thermal layering Occurs when the gases produced by fire stratify into layers based on their temperatures. The hottest gases rise to the highest point in the compartment or room, and the gases with the lowest temperatures stay closer to the bottom of the container, or nearer the floor.

Thick water An additive designed to provide an insulating barrier on the surface of a solid fuel.

Third rail (wire) The rail used in subways and the wire for overhead electric train power systems to provide electrical power to the train.

Time lag The time it takes the moisture content of fuel and the surrounding air to equalize.

Transitional mode The critical process of shifting from the offensive to the defensive mode or from the defensive to the offensive mode. If not conducted safely, firefighters can be endangered as supporting hose lines and protected positions are being changed.

Transom A small window located in an older building, such as a hotel, at the top of the ceiling, generally located over the room entrance to allow heated air to circulate into the room as most were not equipped with heating units in the room.

Ultrafine water mist Water is dispensed under high pressure through very fine nozzle outlets creating nearly micro-sized droplets. This fine mist exhibits high-energy absorption behavior because of the high vaporization rate of the fine water particles (the surface area of the droplet exposed to the heat). Additionally, the gaslike dispersion of the mist acts like a total flooding agent as the fine mist is dispersed throughout the enclosure.

Unified command (UC) The structure used to manage emergency incidents involving multiple jurisdictions or multiple response agencies that have responsibility for control of the incident.

Upper explosive limit (UEL) The highest percentage of a substance in air that will ignite.

Utility chase A channel used for electrical, telephone, and plumbing lines and pipes for the various services in the building. They provide a channel for the movement of fire and heated fire gases.

Vapor density (VD) The mass of the vapor divided by the volume it fills.

Vapor pressure The pressure placed on the inside of a closed container by the vapor in the space above the liquid which the container holds.

Ventilation A systematic process to enhance the removal of smoke and fire by-products to allow the entry of cooler air and facilitate rescue and fire fighting operations.

Viscous water Thickening agents that are added to water to thicken it so it will cling to the surface of the fuel. Because it is thicker, it also provides a more continuous coating over the fuel surface and absorbs more heat.

Weapons of mass destruction (WMD) The term contains the use of hazardous materials, nuclear radiation, biological agents, chemicals, explosives, and cyber and agriculture agents. All are designed to kill or seriously injure numerous people.

Wet water Water with a wetting agent added. The wetting agent reduces the surface tension of the water allowing it to flow and spread better. The improved flowing characteristic allows the water to penetrate tightly packed goods better than water not treated.

White phosphorous This solid material is more dangerous than red phosphorus because of its ready oxidation and spontaneous ignition when exposed to air. It may be stored under water or oil.

Wildland/urban interface The line, area, or zone where structures and other human developments intermingle with undeveloped wildland or vegetative fuels.

Windward side The side from which the wind is blowing. See also *leeward side*.

INDEX

Search and rescue, in high-rise buildings, 174
Securing a building, 152–53
Self-contained breathing apparatus (SCBA), 46, 49
Self-oxidizing materials, 80, 98
Self-reactive materials, 44
Sewer pipes, for water removal, 151
Siamese connection, 137, 155
Side impact curtain (SIC), 210, 237
Single-story family dwelling, 120
Size-ups
  apparatus and staffing, 110
  area and height, 116, 118
  attack modes for, 104–05
  auxiliary appliances, 112–13
  construction type, 105–10
  defined, 102, 130
  exposures, 116
  at incident scene, 105–18
  introduction, 105
  life hazard, 111
  location and extent of fire, 118
  occupancy or use, 110
  postincident, 138–43
  RECEO-VS, 105
  special concerns, 118
  street conditions, 113
  terrain, 111–12
  time, 118
  water supply, 112
  weather, 114–16
  wild land fires, 185
Skylights, 144
Slippery water, 82, 98
Slope, fire spread and, 188–89
Smoke detectors, 13
Smoke ejectors, 146, 155
Smoldering (glowing) combustion, 54–57, 74
Sodium, 96
Sodium hydride, 96
Sodium nitrate, 43
Sodium peroxide, 43

Soffits, 110, 113, 130
Solids, 24, 38
Solubility, 33–34, 49
Sounding a roof, 144, 155
Specific gravity, 34–35, 49
Spontaneous combustion, 54, 74
Spot fires, 188, 194, 205
Spray water applications, 83–85
Sprinkler systems, 13, 172–73
Stack effect, 66, 74, 165, 176
Staffing, during size-up, 110
Stairways, for water removal, 151
Stairwells
  in high-rise buildings, 168–69
  ventilation and, 144
Stairwell support procedure, 170, 176
Standard operating procedure (SOP), 110, 114, 130
Standpipes, 113, 115, 116, 130, 171–72
State emergency response committee (SERC), 247–48, 260
Stoichiometric, 57, 74
Stoichiometry of reaction, 57, 74
Straight steam water applications, 83
Stratification location, 165, 166, 176
Street conditions, during size-up, 113
Strip malls, 122
Strontium, 96
Sublimation, 38, 49
Subway rail vehicles, 224–25
Superfund Amendment and Reauthorization Act (SARA) (1986), 240–41
Supplemental restraint system (SRS), 210, 237
Surfactant, 17, 18, 20
Swamper, 197, 205
Symbols, of elements, 28

T

Tandem action, 202
Tanker ships, 235